高等教育"十四五"规划教材·无人机应用技术

无人机专业英语

（第 2 版）

主　　编　刘肩山

副主编　蒋焕新

编　　者　刘肩山　蒋焕新　谢志明

　　　　　唐　毅　吴　轮

U0382050

西北工业大学出版社

西安

【内容简介】 本书是根据无人机应用技术专业的发展需求并结合无人机行业发展实际编写而成的,主要内容包括无人机职业认知、遥控器使用、无人机组装与调试、无人机飞行基础知识、无人机维护与保养、无人机任务规划等,旨在使读者通过研读的方式掌握无人机应用技术相关专业英文词汇和无人机装调及飞行基础知识。

本书可作为高等院校无人机应用技术专业"专业英语"课程的教材或教辅材料,也可作为企事业无人机工程技术人员的培训资料。

图书在版编目(CIP)数据

无人机专业英语 / 刘肩山主编. —2 版.— 西安 :
西北工业大学出版社,2023.8(2025.1重印)
ISBN 978 - 7 - 5612 - 8714 - 9

Ⅰ.①无… Ⅱ.①刘… Ⅲ.①无人驾驶飞机-英语
Ⅳ.①V279

中国国家版本馆 CIP 数据核字(2023)第 080306 号

WURENJI ZHUANYE YINGYU
无人机专业英语
刘肩山　主编

责任编辑:胡莉巾	策划编辑:杨　军	
责任校对:朱辰浩	装帧设计:董晓伟	
出版发行:西北工业大学出版社		
通信地址:西安市友谊西路 127 号	邮编:710072	
电　　话:(029)88493844,88491757		
网　　址:www.nwpup.com		
印 刷 者:西安五星印刷有限公司		
开　　本:787 mm×1 092 mm	1/16	
印　　张:12.5		
字　　数:328 千字		
版　　次:2018 年 8 月第 1 版	2023 年 8 月第 2 版	2025 年 1 月第 3 次印刷
书　　号:978 - 7 - 5612 - 8714 - 9		
定　　价:49.00 元		

如有印装问题请与出版社联系调换

第 2 版前言

为了进一步贴合无人机产业人才需求,也为了探索"岗课赛证"融通育人模式,我们按照《国家职业教育改革实施方案》和《职业教育提质培优行动计划(2020—2023 年)》的总体要求对《无人机专业英语》(第 1 版)进行了修订。

依据《国家职业技能标准 无人机装调检修工(2021 年版)》《无人机检测与维护职业技能等级标准(2021 年 1.0 版)》《无人机组装与调试职业技能等级标准(2021 年 1.0 版)》《民用无人机驾驶员管理规定》和《民用无人机驾驶员实践考试标准》,并对接无人机行业真实工作任务和职业技能要求,本次具体修订内容包括:增加了无人机维护保养内容,其中对接了近几年国内外无人机大赛中无人机系统常见故障及排除项目;增加了固定翼无人机的组装与调试、无人机模拟飞行操控和真机飞行操控训练等内容;删除了第 1 版的第 1、3~6、11 和 12 章内容,同时对第 1 版的第 10 章内容进行了重新编写;在第 1 版的第 9 章中,增加了完整的任务规划实操内容。另外,对与无人机应用技术专业就业岗位息息相关的无人机组装与调试、无人机飞行操控、无人机检测与维护等知识体系进行解构,以任务为驱动,对内容进行整合,包括将第 1 版的第 13 章内容重构为本版第一个项目的 3 个任务;将第 1 版的第 8 章内容重构为本版第二个项目的 8 个任务;将第 1 版的第 7 章、重新编写后的第 10 章及新增的固定翼无人机装调内容整合为本版第三个项目的 3 个任务;将第 1 版的第 2 章、新增的模拟飞行和真机飞行训练内容整合为本版第四个项目的 3 个任务;将新增的无人机维护内容整合为本版第五个项目的 6 个任务;将修订的第 1 版第 9 章内容整合为本版的第六个项目。这样,教材内容重构为无人机职业认知、遥控器使用、无人机组装与调试、无人机飞行基础知识、无人机维护与保养及无人机任务规划等六大项目,对每个项目都进行了任务描述和任务分析,并列出了所需知识点和技能点。

本书由长沙航空职业技术学院刘肩山任主编,长沙航空职业技术学院蒋焕新任副主编。具体编写分工如下:刘肩山、谢志明和唐毅编写了项目一至四和项目六;蒋焕新和吴轮编写了项目五。

在编写本书的过程中参考了大量相关资料,在此对其作者一并表示感谢!

由于水平有限,书中难免存在欠妥之处,恳请广大读者批评指正。

编 者
2022 年 10 月

第1版前言

近年来,随着科学技术的发展,无人机在民用领域的应用越来越广阔,从无人机航拍、植保、电力巡检、消防、物流和航测等基础应用,到现在兴起的各种无人机大赛等极限运动,无人机以各种形式活跃于各行各业。正是由于民用无人机市场的迅速发展和不断升温,相关的人才培养成为高校必然要涉及的领域,目前国内外的许多高校已经纷纷开始设立无人机应用技术专业。为了培养学生的国际化视野,提高学生语言应用能力,学好无人机专业英语,在当前乃至未来很长一段时间都具有重要的意义。

在无人机领域,美国优势技术突出,欧洲各国以及日本紧随其后,中国发展迅速。无人机领域从业人员必须掌握相应的专业技术词汇,才能阅读并真正理解英文技术资料,从而了解最新的无人机技术和无人机发展趋势,这对从事无人机研发、制造、维护及使用等相关领域的人员具有重要的作用。

本书根据高职院校开设的无人机应用技术专业需求,突出职业特色,结合现阶段无人机人才需求实际,并按照《民用无人机驾驶员管理规定》及民用无人机驾驶员考试标准的要求,重点围绕无人机组装与调试过程中用到的基本知识和技能,精心挑选、编写了13单元的内容,包括无人机概述、四轴无人机介绍、无人机导航定位技术、无人机组装与调试、无人植保机使用以及无人机驾驶员考证相关的知识,尽可能地涵盖无人机应用技术领域的各个方面,使得读者在精读本书内容后,能达到如下目的:掌握无人机应用技术专业英文词汇;能够阅读英文说明书、技术手册和相关文献;能够更多地了解无人机发展现状及前景;对无人机法规及驾驶员考证也能进行全面的了解。

在本书编写中,以无人机合格证要求的航空理论知识为主线,贯穿必需的知识点,注重基本知识、基本技能和基本方法的介绍,突出无人机组装与调试实际应用,并辅之无人机行业应用和法规等内容的介绍,兼顾知识的系统性和逻辑性,力求结构合理,图例丰富。为了便于教学,本书配有相应的教学课件。

本书由长沙航空职业技术学院刘肩山和成都航空职业技术学院张伟瑞主编。具体分工如下:刘肩山编写第 1~3,6~9 和 11~13 章;张伟瑞编写第 4,5 和 10 章。全书由于坤林和蒋焕新审校。西北工业大学航空学院刘贞报教授对本书进行了审阅,并提出建设性意见,使本书更加完善。

本书是西北工业大学出版社联合全国无人机教育联盟开发的一套无人机应用技术系列教材之一。本书可作为高等院校无人机应用技术专业"无人机专业英语"课程的教材或教学辅导用书,也可用于企事业单位无人机工程技术人员的培训教材。

在编写本书的过程参考了大量无人机相关资料,同时也参考了互联网上的相关资料,在此一并表示感谢!

由于学识水平有限,书中难免有不足或欠妥之处,恳请同行和广大读者批评指正。

编　者

2018 年 5 月

Contents

Project 1   What Is a UAS Pilot Job Like?

【Description】 This project is about the UAS pilot occupation, students are required to search for information through the internet and library or acquire first-hand data from friends and relatives. The project aims to make students of UAV specialty have a clear understanding of the UAS pilot, helping to guide personal occupation career planning well.

【Analysis】 Students are divided into several groups and required to hold a discussion about the UAS pilot occupation from three aspects mainly, they are the UAS pilot job, occupational pre-requisite and work activities.

【Related knowledge and skills】 Introduction to UAV applications; introduction to requirements set forth for Drone Pilot; introduction to occupation competency of Drone Pilot.

Read and translate some long and difficult sentences.

Mission 1   Occupation Overview

【Objective】 Students are required to discuss about the UAV applications and the market trends. The objective of this mission is mastering the key vocabularies about the UAV job, understanding the key sentences and knowing the main UAV occupations.

【Analysis】 UAVs are applied in many fields; we can start from illustrating with examples of UAV applications and make an analysis of market trends with data from websites.

【Knowledge preparation】 UAV terminology and classification.

The drone industry is one of the newest industries in the world. Since its infancy in the past 10 years, the operation of drones has always been with the military and researchers. The landscape has dramatically changed with the advancement of miniaturizing components and lower costs. According to industry estimates, the drone business can generate billions for the world economy and create thousands of new jobs over the next 10 years.

This has created a new opportunity for a new breed of drone enthusiasts. Many are hobbyists that we would categorize as recreational drone operators. Such activities like building and flying DIY drones and participating in drone racing or drone soccer has taken traction.

The natural progression of entering the amateur market as well as undertaking a lot of aerial photography to hone in on their flying skill is now very popular. Those who do so have also found that there is some remuneration involved and finally entering the commercial market is the goal of some pilots who see a career as a Drone Pilot.

The demand for Drone Pilots has been identified in nine industries namely:

(1) Agriculture (see Fig. 1 - 1 - 1), Aquaculture, Silviculture & Viticulture.

(2) Security Services.

(3) Infrastructure.

(4) Health, Emergency Services & Disaster Recovery.

(5) Environmental Management.

(6) Media Communications.

(7) Urban Planning, Real Estate, Architecture & Engineering.

(8) Business & Commerce.

(9) Recreation & Entertainment.

Fig. 1 - 1 - 1   UAV for plant protection

Drone pilots can be a well-paying profession. It is a rapidly expanding vocation that is playing a larger and larger role in several industries. If you are interested in making this jump, there is one question you will need to ask first: what jobs are available for skilled drone pilots?

The drone market is a train that continues to build momentum and shows no signs of slowing down. With thousands of expert drone pilots, the business is now as flourishing as ever and develops steadily about diversity and quantity.

The development of the job market has made a career as a professional drone pilot a rather attractive prospect for many people. But there are a few areas where professional pilots are making powerful effects.

The following sections highlight some of the specialties where drone pilots have made prosperous inroads. If you are considering becoming a professional pilot, below are a few of the industries you may want to check into first.

Here's a quick rundown of the hottest drone jobs.

1.1.1   Drone 3D Modeler

Creating 3D models and maps using drones is a job with wide applications. Any industry that needs heavy construction like mining, urban planning, and civil engineering may benefit from a 3D terrain model (see Fig. 1 – 1 – 2). A good, comprehensive model can help these firms plan how much excavation has to be performed, the slope and length of streets that will need to be performed, or the mitigating measures for classified geological hazards.

Fig. 1 – 1 – 2   3D terrain model

There are two major methods for creating 3D models using drones. The easier and more commonly practiced one is photogrammetry (see Fig. 1 – 1 – 3), which essentially generates 3D models with a set of overlapping aerial pictures (see Fig. 1 – 1 – 4). Its strength lies in its simplicity—all you will need to have is a drone with a fantastic camera and a distinctive photogrammetry software platform. If you can get software that can help you plan the scope of your survey and make a flight route for your drone, then you can make your life so much simpler.

Fig. 1 – 1 – 3   Photogrammetry

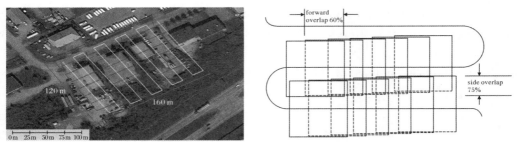

Fig. 1 - 1 - 4   Overlapping aerial pictures

The precision and resolution of photogrammetry can be confined to a 1 to 5 cm variety using RTK (Real Time Kinematic) GPS. If you want something more precise, you will need to use a Light Detection and Ranging (LiDAR) sensor (see Fig. 1 - 1 - 5). This sort of sensor is significantly more expensive but can produce models with accuracy levels several orders of magnitude compared to photogrammetry.

Fig. 1 - 1 - 5   LiDAR

Regardless, taking a job as a drone 3D mapper or modeler is easily among the very productive you can get as an expert drone pilot. There are so few pilots who may supply this service, which means you could charge premium prices.

1.1.2   Drone Photographer and Filmmaker

Let's lump all aerial filmmaking and photography jobs under one thing, just because explaining all kinds of tasks under this category would be long enough to make another list. Being among the most enticing fields for drone pilots, drone filmmaking and photography (see Fig. 1 - 1 - 6) have the largest market and the greatest number of pilots that offer the service.

There are likely more than a dozen ways for drone photography to be rewarding. You can capture photos and sell them online, provide event coverage, offer your services to print advertisers, or take aerial pictures for realtors. It's a simple field to get into, mostly because many photographers have expert-level skills.

Fig. 1 – 1 – 6   Drone photographer

Drone filmmaking is a lucrative field that's fantastic for people who already have filmmaking experience. Aerial videos taken by drones have become a standard element of contemporary film. If you can hook up with a movie producer, then you've got an opportunity to earn hundreds to thousands of RMBs capturing a couple of minute's values of aerial footage.

If you can capture aerial photographs of news-worthy events, then there is also the possibility it may capture the eyes of a news service or a documentary producer. Even private corporations typically employ the help of a drone filmmaker to take corporate videos. Additionally, there are the typical advertising businesses and private events which are searching for aerial video coverage.

Cinematographers used the technology for James Bond to chase scenes in 2012 and other blockbusters like *The Wolf of Wall Street*. In 2015, filmmakers created a mixed scene in Jurassic Earth, where a piloted drone captured a bunch of individuals responding to a dinosaur attack.

In *Tinseltown*, drone cinematography is a cost-effective resource used alongside conventional gear, like dollies and jibs. For the time being, they have limited battery life and are only allowed to fly to a specific height, which makes them impractical for some applications.

However, when drone usage makes sense, filmmakers save substantial costs. For example, a helicopter may cost $20,000 – $40,000 for a 10-hour shoot, while filmmakers can usually capture the same scene using a drone at a price of just $4,500 – $13,000 daily.

The drone filmmaking marketplace may not be as big, but drone filmmakers can generally charge higher prices than drone photographers.

1.1.3   Power Line Builder

Power lines have been damaged over time and require regular maintenance to stay in working condition. The challenge lies in inspecting miles' worth of electricity lines merely to search for damaged segments, which can be quite short. Needing to do this manually requires quite a long time, and of course, the risk that power line inspection teams become exposed to if they need to work at heights.

Drones take away this threat and can even do the job quicker. Since power lines generally run through narrow corridors, drones can quickly fly and pay several miles on a single battery bicycle. There is also the matter of determining damaged sections—a problem solved with a special device called a thermal camera.

Thermal imaging (see Fig. 1 - 1 - 7) depends on the infrared energy radiated by all things. This infrared energy is stronger when things emit heat. With this principle, the heat accumulation due to interrupted current flow in degraded power lines is easily identified.

Fig. 1 - 1 - 7   Thermal imaging of power line

By flying a drone armed with a thermal camera to inspect power lines for damage, the job that would normally require a few weeks may be completed in many hours. It is also safer and logistically easier—all you will need is a crew composed of the visual observer, drone operator, and maybe someone to drive a motor vehicle. This can be a very rewarding endeavor for a drone pilot as it would be largely utility companies who will ask for this kind of service.

1.1.4 Search and Rescue Team Member

Law enforcement agencies and emergency responders globally have taken to using drones to help strengthen their capacities. For situations that need search and rescue, a drone is an ideal tool. It can cover lots of ground quickly, provides a bird's eye perspective of the research area, and may be deployed at a moment's notice.

The same thermal camera used for reviewing industrial equipment and power lines may prove valuable in search and rescue works. Unlike normal eyesight, which could be obscured by vegetation or darkness, a thermal camera can spot high-temperature aberrations even at night or under a canopy cover.

Some drones can be outfitted with a speaker, which is used to draw out missing individuals. In this respect, the drone behaves as a mobile spotlight, playing the same role that a helicopter could under different conditions. Of course, flying a drone is easier and a lot less expensive than flying a helicopter.

Law enforcement agencies would probably prefer to have an in-house drone pilot one of their ranks, which will permit them to participate in drone-aided search and rescue at a minute's notice. Should your neighborhood law enforcement agencies need an officer to serve as a drone pilot, you might still offer your services to give their own department training.

1.1.5　Drone Flight Instructor

Concerning providing training, how about being a full-time drone flight instructor? For those that have a knack for teaching and speaking to people, this might sound like an excellent idea. However, you may ask if there is another certification you want to earn before you may be a drone flight instructor.

The short answer is yes, and you will find there are certification standards that the AOPA [Aircraft Owners and Pilots Association of China (APOA-China)] grants for people who wish to educate drone flight.

Most organizations have set their standards for drone flight instructors. For example, one site that provides online courses may accept only people with a manned pilot's license. Others may take a minimum number of years of expertise in droning flight, especially in industrial operations.

Of course, Experience is a given if you're looking to be a drone flight instructor. Take those flight hours, and you might be good enough to educate others in a couple of years.

Words & Phrases

infancy ['ɪnfənsi] n. 初期
hone in on [həʊn ɪn ɒn] 磨炼
remuneration [rɪˌmjuːnəˈreɪʃn] n. 薪水，报酬
aquaculture ['aːkwəkʌltʃər] n. 水产养殖业
silviculture ['sɪlvɪˌkʌltʃə] n. 森林学，造林学
viticulture ['vɪtɪkʌltʃər] n. 葡萄栽培

infrastructure [ˈɪnfrəstrʌktʃər] *n.* 基础建设

lucrative [ˈluːkrətɪv] *adj.* 赚大钱的

rundown [ˈrʌndaʊn] *n.* 介绍，描述

photogrammetry [ˌfəʊtəʊˈɡræmɪtrɪ] *n.* 摄影测量学

overlap [ˌəʊvəˈlæp] *n.* 重叠

resolution [ˌrezəˈluːʃn] *n.* 分辨率

LiDAR [ˈlaɪdɑː] *n.* 激光雷达

photographer [fəˈtɒɡrəfə(r)] *n.* 摄影师

power line [ˈpaʊə laɪn] 输电线

infrared [ˌɪnfrəˈred] *adj.* 红外线的

enforcement [ɪnˈfɔːsmənt] *n.* 执行，实施

canopy cover [ˈkænəpi ˈkʌvə(r)] 冠层覆盖

Exercises and Thinking

1. Translate the following sentences.

1) The natural progression of entering the amateur market as well as undertaking a lot of aerial photography to hone in on their flying skill is now very popular.

2) Any industry that needs heavy construction like mining, urban planning, and civil engineering may benefit from a 3D terrain model.

3) Law enforcement agencies would probably prefer to have an in-house drone pilot one of their ranks, which will permit them to participate in drone-aided search and rescue at a minute's notice.

2. Read the following recruitment information and answer what responsibilities and requirements the job need.

Responsibilities:

1) Operate UAV to complete aerial photography, surveying and mapping, power-line inspection projects;

2) The daily maintenance, assembling and adjustment of UAV;

3) Make flight plans of flight area and arrange the material that flight needed;

4) Organize the flight process and do on-the-spot investigation;

5) Sort the flight results and submit them to customers.

Requirements:

1) College diploma or above; major in Electronics, Mechanics, Telecommunications, or related industry.

2) Be enthusiastic about UAV industry, at least 2 years of experience of operating UAV, owe civil UAS pilot certificate of AOPA and Driver license (C2).

3) Be experienced in UAV mapping, powerline inspections and have long secure flight record.

4）Good communication skills, reliable and Integrity, clear thinking, be able to deal with and coordinate the outfield emergencies.

5）Be able to adapt to business trip and work hard.

3.Read the following passages and answer questions.

Looking for a Drone Pilot Job

Here you will find out how to get a drone pilot job. A new wave of pilot jobs in the drone enterprise may grow beyond the typical tech hotbeds, which are attached to research institutes in rich regions and to delivery and logistics hubs, large industrial installations, and construction sites.

Why it matters: it's not clear how many jobs drones will offer, but where new jobs are and who will be equipped to do them will help determine who will benefit from the Drone Age.

The big picture: thousands of people already use drones to design, maintain, manufacture, and fly. However, the industry is still in its early phases, slowed in part by regulatory requirements.

Experts' predicts for the future of this work vary broadly, from a sustained steady trickle of new jobs to a broader deluge.

Drone-related hiring appears focused in big cities not just in traditional tech hubs. It is so easy to see the explosive nature of this.

Every business you look at will possibly be touched by drones. Anywhere you'd go and wait in a line and not be happy about waiting in that line, simply assume there will be a delivery option available.

In data that ZipRecruiter shared, job openings ranged from the prominent (developer, pilot) to the unexpected (lifeguard, camp counselor).

However, in life, it feels like it is moving a little slower. Drone training courses are in demand, but not overflowing. One thing holding back the uptake is companies' ignorance of how to incorporate drones.

Many companies consider drones very easy tool, but they do not consider regulations and the maintenance program.

The big question: When the administration ultimately allows most pilots to operate drones remotely—surpassing their line of sight—will operations be combined into a central location?

Instead of having pilots remaining at every logistics hub, a shipping company like Amazon or SF could place them all in an office building somewhere and have them manage flights globally.

That—also increasingly independent drones—might chip away at the vision of universal drone jobs.

While there is no denying the steady need for UAS services, the supply of pilots is more or less keeping up with this requirement. This has resulted in downward pricing pressure and an intense environment competitive on drone solutions. Therefore, UAS pilots need to have a solid business plan in place.

We give several tips to break into the drone industry and then progressively build up a steady and profitable drone market. Which drone jobs should novices go after? Should you find drone jobs through websites? Which are the most profitable opportunities to get a drone pilot job?

Questions:

1)List the main applications of UAV.

2)Do you expect to be a drone pilot? Why?

3)If you own a civilian drone certificate, which kind of drone pilot job do you expect to get?

Mission 2　Occupational Pre-requisite

【Objective】Students are required to study regulations on http://www. aopa. org. cn/ and be familiar with the requirements of being a drone pilot. The aim of this mission is mastering the key vocabularies about AOPA exam, understanding the key sentences and knowing the procedures of AOPA exam.

【Analysis】It's not necessarily to have a license to be a drone pilot, but your operations with UAVs may be illegal and pose a threat to others. Not everyone can fly UAVs legally, so we should have intimate knowledge of occupational pre-requisite. And the best way is searching information on the AOPA website which gives the detailed information.

【Knowledge preparation】UAV classification; flight mode.

Candidates for Drone Pilot should fulfil all requirements set by the relevant statutory bodies.

The requirements set forth for Drone Pilot are as follows:

(1)Basic requirements;

(2)Experience requirements;

(3)Oral, practical and written tests.

1.2.1　Basic Requirements

(1)Be at least 16 years old;

(2)Be able to read, speak, write (exceptions may be made if the person is unable to meet one of these requirements for a medical reason, such as hearing impairment);

(3)Medically fit to safely operate a small UAS;

(4)No criminal record certification;

(5)Receive at least junior middle school education.

1.2.2   Experience Requirements

You can attend one of training organizations that granted by AOPA or UTC (Unmanned Aerial Systems Training Center). These organizations offer courses covering theoretical knowledge and practical operation.

You need a high school diploma to get in to most training organizations. The training generally lasts 1 months. When you finish all courses, you are qualified to take AOPA or UTC exams.

1.2.3   Oral, Practical and Written Tests

This part, we take AOPA exams for example to depict the procedures of Oral, Practical and Written Tests.

1. Application process

Schedule an appointment with a Knowledge Testing Center, which administer initial and recurrent UAV knowledge exams.

View the list of Knowledge Testing Centers to find one near you.

Applicants must bring government-issued photo ID to their test.

2. Written test

There are 100 single-choice questions. For WVR pilot, you need to get over 70 points; while for BVLOS pilot, 80+ points are necessary.

Only when you pass written tests, can you attend oral tests.

Pass the initial aeronautical knowledge test-initial knowledge test areas include:

(1)Applicable regulations relating to small unmanned aircraft system rating privileges, limitations, and flight operation.

(2)Airspace classification and operating requirements, and flight restrictions affecting small unmanned aircraft operation.

(3)Aviation weather sources and effects of weather on small unmanned aircraft performance.

(4)Small unmanned aircraft loading and performance

(5)Emergency procedures.

(6)Radio communication procedures.

(7)Determining the performance of small unmanned aircraft.

(8)Physiological effects of drugs and alcohol.

(9)Aeronautical decision-making and judgment.

(10)Maintenance and preflight inspection procedures.

3.Oral test

In oral test, you are given 10 single-choice questions, and the passing grade is 7. When you pass the oral test, you can go on with the practical test.

4.Practical test standards

(1)Fixed-wing plane：

WVR pilot can use stabilize mode (flight controller internal loop control).

1)Take off (wheeled/Catapult/Hand throwing).

2)Engine failure stimulation at the 3rd side, Simulated touchdown height is less than 5 m.

3)Land or recycle at designated reign.

BVLOS pilot can use stabilize mode (flight controller internal loop control).

1)Take off (wheeled/Catapult/Hand throwing).

2)Horizontal "8" track, diameter of both cycles is 50 m.

3)Engine failure stimulation at the 3rd side, Simulated touchdown height is less than 5 m.

4)Ground station route planning, modify height and position of waypoint, emergency operation.

5)Land or recycle at designated reign.

(2)Rotor：

WVR pilot (helicopter/multirotor)can use GPS mode (both internal and outside loop of flight controller participate in control).

1)Hover.

2)Horizontal 360° track at low speed.

3)Horizontal "8" track, diameter of both cycles are 6 m.

4)Land at designated point.

BVLOS pilot (helicopter/multirotor) can use stabilize mode (flight controller internal loop participates in control; flight controller cannot hover at fixed point).

1)Hover.

2)Horizontal 360° track at low speed.

3)Horizontal "8" track, diameter of both cycles is 6 m.

4)Ground station route planning, modify height and position of waypoint, emergency operation.

5)Land at designated point.

After you have passed Oral, Practical and Written Tests, congratulations! You can get a civil UAS pilot certificate (see Fig. 1 - 2 - 1).

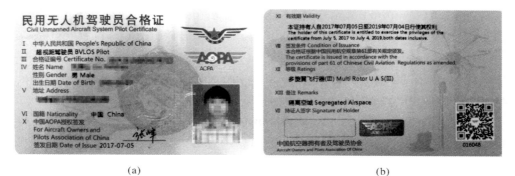

(a)　　　　　　　　　　　　　　　　　　　(b)

Fig. 1 − 2 − 1　AOPA certificate

(a)front side;(b)back side

(3)Pilot certificate requirements:

1)Must be easily accessible by the remote pilot during all UAS operations.

2)Valid for 2 years-certificate holders must pass a recurrent knowledge test every two years.

Words & Phrases

WVR＝Within Visual ['vɪʒuəl] Range *abbr.* 在视距范围内

BVLOS＝Beyond [bɪ'jɒnd] Visual ['vɪʒuəl] Line of Sight 视距外

privilege ['prɪvəlɪdʒ] *n.* 权限

aeronautical [ˌeərə'nɔːtɪkl] *adj.* 航空的

Exercises and Thinking

1. Translate the following sentences.

1) These organizations offer courses covering theoretical knowledge and practical operation.

2) Applicable regulations relating to small unmanned aircraft system rating privileges, limitations, and flight operation.

3) Engine failure stimulation at the 3rd side, Simulated touchdown height is less than 5m.

2. Answer the questions in your own words.

1) How do I get a civil UAS pilot certificate?

2) What's the practical test standards for WVR pilot of rotor?

3) What's the basic requirements for Drone Pilot.

3. Read the following passages and answer questions.

How Do I Become a Certified Drone Pilot in America?

Under Part 107, which the FAA announced on June 21st, 2016 and implemented on August 29th, 2016, commercial drone operators are required to:

(1) Pass an initial aeronautical knowledge test at one of around 700 FAA-approved knowledge testing centers across the United States (this list last updated May 2017). Be vetted by the Transportation Security Administration.

(2) Obtain a Remote Pilot Certificate with a small UAS rating (like existing pilot airman certificates, never expires).

(3) Pass a recurrent aeronautical knowledge test every 24 months.

(4) Be at least 16 years old.

(5) Make available to the FAA, upon request, the small UAS for inspection or testing, and any associated documents/records required to be kept under the proposed rule.

(6) Report an accident to the FAA within 10 days of any operation that results in injury or property damage over $500.

(7) Conduct a preflight inspection, to include specific aircraft and control station systems checks, to ensure the small UAS is safe for operation.

For those who need to operate outside the flight and mission parameters of Part 107, you'll need to gain additional permission from the FAA through a waiver process. Things like flying at night, operating beyond visual-line-of-sight (BVLOS), etc.

Questions:

1) What's the requirements to be a commercial drone operator?

2) If you need to operate outside the flight and mission parameters of Part 107, what do you need?

Mission 3 Work Conditions and Competency Profile

【Objective】Students are required to do survey about drone pilot work conditions and what kind of competency needed to become a drone pilot. The aim of this mission is mastering the key vocabularies about competency, understanding the key sentences and knowing the competency of drone pilot.

【Analysis】Each job has its own characteristics, and what people really concern are work conditions and competency. We can search on the internet to acquire information, especially the recruitment announcement.

【Knowledge preparation】UAV flight principles; the structure of UAV.

Drone Piloting personnel could be employed by various sectors and industries either local or foreign companies. They can earn attractive salaries and may be required to travel frequently depending on selected industries. Other related industries which a Drone Piloting with respect for employment opportunities are:

(1)Oil and gas industry.

(2)Port services industry.

(3)Plantation industry.

(4)Defence and security industry.

(5)Government agencies.

1.3.1　Work Conditions

Being a drone pilot is very exciting and demanding. You will be exposed to many industries in your line of work. Every project has its own challenges and the drone pilot will be faced with different conditions and demands which would require some degree of initiative, resourcefulness and proactive action.

The drone pilot is exposed to working outdoor and in remote areas where necessary but spends time as well on work preparations which include pre-flight planning and general maintenance.

When in the field, the drone pilot must prioritize safety at all time ensuring being alert to weather and environment conditions.

1.3.2　Competency Profile

1. Drone Pre-flight Preparation

Drone Pre-flight Preparation describes the knowhow of the technician to setup the equipment & devices before flying operation.

The person who is competent in this competency unit shall be able to charge drone battery, format storage card, propellers, check weather and environment condition, install drone sensors, record pre-flight log, setup Ground Control Station (GCS), check compass calibration, and locate take-off/landing zone.

The outcome of this competency unit is to enable the readiness of the rotary drone for flight operation as required by project brief.

2. Drone Flying Operation

Drone Flying Operation describes the responsibility of preparing and setup devices and equipment before drone pilot engage in flight operation.

The person who is competent in this competency unit shall be able to check drone abnormalities, check thrust/take-off drone, manoeuvre drone, engage flying mission, execute Emergency Recovery Procedures (ERP) and prepare drone landing.

The outcome of this competency unit is to ensure drone operate smoothly during flying mission with all objectives of the mission achieved.

3. Drone Post-flight Operation

Drone Post-flight Operation describes knowhow and responsibility during post drone flying mission.

The person who is competent in this competency unit shall be able to check data collection, inspect drone abnormalities, uninstall drone parts and record post-flight log.

The outcome of this competency unit is to ensure drone data collection from its sensor is recovered and collected safely, drone parts uninstalled properly & flight log recorded.

4. Drone Maintenance

Drone Maintenance describes rotary drone knowhow and responsibility during maintenance coordination.

The person who is competent in this competency unit shall be able to maintain drone battery system, maintain drone motor and avionic sensors, maintain propellers, maintain drone accessories, maintain radio controller, maintain drone frame and maintain firmware.

The outcome of this competency unit is to ensure drone, sensors and accessories coordination for maintenance are done properly as per manufacturer's specifications.

5. Drone Storage Operation

Drone Storage describes the technician responsibility to store equipment and devices in proper arrangement and good condition which are ready to be use whenever the task arises.

The person who is competent in this competency unit shall be able to prepare drone inventory control system, perform drone requisition and acceptance, perform drone parts Incoming Quality Control (IQC) and archive flight log data.

The outcome of this competency unit is to ensure the equipment and devices are systematically stored for the next drone operation.

Words & Phrases

competency ['kɒmpɪtənsi] *n.*能力
manoeuvre[mə'nuːvə(r)] *n.*机动动作
accessory [ək'sesəri] *n.*附件
requisition [ˌrekwɪ'zɪʃn] *n.*正式要求,需要

Exercises and Thinking

1. Translate the following sentences.

1)Every project has its own challenges and the drone pilot will be faced with different conditions and demands which would require some degree of initiative, resourcefulness and proactive action.

2) The outcome of this competency unit is to enable the readiness of the rotary

drone for flight operation as required by project brief.

3) Drone Storage describes the technician responsibility to store equipment and devices in proper arrangement and good condition which are ready to be use whenever the task arises.

2. Answer the questions in your own words.

1) Describe the working conditions of a drone pilot briefly.

2) What should you do to prepare for a flight?

3) Describe the drone post-flight operation briefly.

3. The National Occupational Skills Standard (NOSS) is developed for various occupational areas. Below is a guideline of each NOSS Level as defined by the Department of Skills Development, Ministry of Human Resources. Read the standard and answer questions.

Level 1: Competent in performing a range of varied work activities, most of which are routine and predictable.

Level 2: Competent in performing a significant range of varied work activities, performed in a variety of contexts. Some of the activities are non-routine and required individual responsibility and autonomy.

Level 3: Competent in performing a broad range of varied work activities, performed in a variety of contexts, most of which are complex and non-routine. There is considerable responsibility and autonomy and control or guidance of others is often required.

Level 4: Competent in performing a broad range of complex technical or professional work activities performed in a wide variety of contexts and with a substantial degree of personal responsibility and autonomy. Responsibility for the work of others and allocation of resources is often present.

Level 5: Competent in applying a significant range of fundamental principles and complex techniques across a wide and often unpredictable variety of contexts. Very substantial personal autonomy and often significant responsibility for the work of others and for the allocation of substantial resources features strongly, as do personal accountabilities for analysis, diagnosis, planning, execution and evaluation.

Questions:

1) What's the differences between Level 1 and Level 2?

2) What's the differences between Level 4 and Level 5?

3) Describe the Level 3 using your own words briefly.

【Project Evaluation】

Work activities	Weightage	Scores
Words dictation	15	
Reading and translation	20	
UAV applications	5	
Basic requirements to be drone pilot	5	
Experience requirements to be drone pilot	5	
Procedures of oral，practical and written tests	20	
Competency units	20	
Attitude and attendance	10	

Grade：□Excellence　□Good marks　□Medium level　□Pass

【Project Conclusions】Upon completion of this project，students shall be able to：

1）Understand what is a drone pilot job like；

2）Know UAV uses well；

3）Know how to get a UAS pilot certificate；

4）Get a thorough understanding of competency units of drone pilot.

Project 2　How to Use the Radio Control System?

【Description】 This project will guide you to master the main operation of remote controller, you will need Futaba 14SG to follow the steps and do exercises.

【Analysis】 You should preview the key words in advance and understand them, then listen to the teacher carefully for the explanation of basic operation, figure out the function of each operation, follow the instructions and do a lot of practice.

【Related knowledge and skills】 Introduction to radio controller for drone.

Read and translate some long and difficult sentences; select radio controller for drone.

Mission 1　Link Procedure

【Objective】 Build radio connection between the transmitter and receiver.

【Analysis】 Each transmitter has an individually assigned, unique ID code. In order to start operation, the receiver must be linked with the ID code of the transmitter with which it is being paired. Once the link is made, the ID code is stored in the receiver and no further linking is necessary unless the receiver is to be used with another transmitter. When purchasing additional receivers, this procedure is necessary; otherwise, the receiver will not work.

【Knowledge preparation】 UAV antenna; electromagnetics.

Steps are described as follows.

(1) Place the transmitter and the receiver close to each other within half meter (see Fig. 2 - 1 - 1).

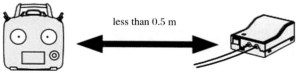

less than 0.5 m

Fig. 2 - 1 - 1　Distance between transmitter and receiver

(2) Turn on the transmitter.

(3) Select [SYSTEM] at the Linkage menu and access the setup screen shown below by touching the RTN button (see Fig. 2 - 1 - 2).

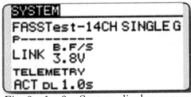

Fig. 2 - 1 - 2　System display screen

(4) When using two receivers on one model, you must change from [SINGLE] to [DUAL].

(5) "F" will be chosen if it is used in France. Others are "G" general.

(6) [LINK] is chosen by scrolling and the RTN button is pushed. The transmitter will emit a chime as it starts the linking process.

(7) When the transmitter starts to chime, power on the receiver. As shown in Fig. 2 - 1 - 3, the receiver should link to the transmitter within about 1 second.

In "Link" mode

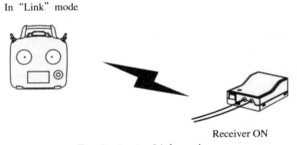

Receiver ON

Fig. 2 - 1 - 3　Link mode

(8) If linking fails, an error message is displayed. Bring the transmitter closer to the receiver and repeat the procedure above from Step(2).

(9) ACT will be chosen if telemetry is used. It is INH when not using it.

This operation is necessary for operating drones. When doing this, please ensure there are no other radio control systems and be far away from electromagnetic interference. Link is required when a system type is changed and linking is required whenever a new model is made. Do not perform the linking operation when the drive motor is connected or the engine is running.

After the linking is done, please cycle receiver power and check that the receiver to be linked is really under the control of the transmitter.

Words & Phrases

transmitter [trænz'mɪtə(r)] n. 发射机

linkage['lɪŋkɪdʒ] n. 连接；联系

chime[tʃaɪm] n. 铃声；(尤指)钟声

telemetry [tə'lemətri] n. 遥测

interference [ˌɪntə'fɪərəns] n. 干扰

Exercises and Thinking

1. Translate the following sentences.

1) In order to start operation, the receiver must be linked with the ID code of the transmitter with which it is being paired.

2) When doing this, please ensure there are no other radio control systems and be far away from electromagnetic interference.

3) Do not perform the linking operation when the drive motor is connected or the engine is running.

2. Read the following passages and answer questions.

If there are many FASSTest systems turned on around your receiver, it might not link to your transmitter. In this case, even if the receiver's LED stays solid green, unfortunately the receiver might have established a link to one of other transmitters. This is very dangerous if you do not notice this situation. In order to avoid the problem, we strongly recommend you to doublecheck whether your receiver is really under control by your transmitter by giving the stick input and then checking the servo response.

When you use two receivers, please be sure to setup a "primary" and "secondary" in the "dual" mode.

Since two sets of receivers cannot be individually recognized without using a "primary" and "secondary" setup, it is impossible to receive telemetry data correctly.

You must link one receiver at a time. If both power supplies to the receivers are switched on simultaneously, data is received incorrectly by the transmitter.

If a dual receiver function is used, in order to receive sensor information correctly by both sets, telemetry data will be slower compared to a single receiver setup.

Link is required when a system type is changed and whenever a new model is made.

Questions:

1) When linking to the transmitter, what should be considered?

2) To avoid linking to the wrong transmitter, what should we do?

3) In which situation is the link procedure necessary?

Mission 2　H/W Setting

【Objective】 Set the direction of the sticks, switches, trimmer levers and knobs; change the stick mode of transmitter; calibrate the stick of transmitter.

【Analysis】 This mission can reverse the hardware of transmitter and select the appropriate stick mode for pilot. It's very useful when you want to change the acting direction of control surfaces or you want to operate the drone with American/Japanese hand mode.

【Knowledge preparation】The components of transmitter; American/Japanese hand mode.

As shown in Fig. 2 – 2 – 1, the basic H/W Setting procedures are depicted below.

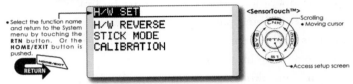

Fig. 2 – 2 – 1   H/W Setting procedures

2.2.1   H/W Reverse

This function reverses the operation direction of the sticks, switches, trimmer levers, and knobs.

(1)Select [H/W REVERSE] (see Fig. 2 – 2 – 2) and access the setup screen shown below by touching the RTN button.

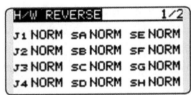

Fig. 2 – 2 – 2   H/W REVERSE screen

(2)Move the cursor to the item corresponding to the H/W you want to reverse and touch the RTN button to switch to the data input mode.

(3)Select the mode by scrolling the touch sensor. The display blinks. When the RTN button is touched, the operation direction is reversed.

"NORM": Normal operation direction.

"REV": Operation direction is reversed.

2.2.2   Stick Mode

This function changes the stick mode of transmitter. J1 – J4 stick correction can be performed.

Select [STICK MODE] and access the setup screen shown below by touching the RTN button (see Fig. 2 – 2 – 3).

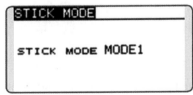

Fig. 2 – 2 – 3   STICK MODE screen

(1)Move the cursor to the "STICK MODE" item and touch the RTN button to switch to the data input mode.

(2)Select the mode from 4 different stick modes (see Fig. 2 – 2 – 4). The display blinks. When the RTN button is touched, the stick mode is changed.

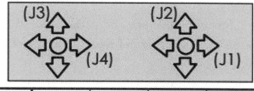

Mode	J1	J2	J3	J4
1	Aileron	Throttle	Elevator	Rudder
2	Aileron	Elevator	Throttle	Rudder
3	Rudder	Throttle	Elevator	Aileron
4	Rudder	Elevator	Throttle	Aileron

Fig. 2 – 2 – 4　Stick mode

2.2.3　Stick Calibration

(1) Select [CALIBRATION] and access the setup screen shown below by touching the RTN button (see Fig. 2 – 2 – 5).

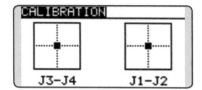

Fig. 2 – 2 – 5　Stick calibration screen

(2) Move the cursor to the J3 – J4 button and touch the RTN button.

(3) Move the J3 or J4 sticks to the neutral position and press the RTN button for one second.

(4) Set the J3 and J4 sticks fully to the bottom right (see Fig. 2 – 2 – 6) and wait until the buzzer sounds.

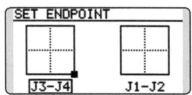

Fig. 2 – 2 – 6　Set the bottom right endpoint

(5) Set the J3 and J4 sticks fully to the top left (see Fig. 2 – 2 – 7) and wait until the buzzer sounds.

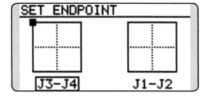

Fig. 2 – 2 – 7　Set the top left endpoint

(6) The above completes the correction operation. Operate and check if stick correction was performed normally.

The "H/W reverse" function reverses the actual operation signal, but does not change the display indicators. Use the Normal mode as long as there is no special reason to use the Reverse mode.

The "stick mode" function will not change the throttle ratchet, etc. Those are mechanical changes that must be performed by a Futaba service center. After changing the mode, these changes are only applied to new models. It is not applied to an existing model.

Words & Phrases

knob [nɒb] *n.* 旋钮

reverse [rɪ'vɜːs] *adj.* 反向的

neutral ['njuːtrəl] *adj.* 中立的，归中的

Exercises and Thinking

1. Translate the following sentences.

1) This function reverses the operation direction of the sticks, switches, trimmer levers, and knobs.

2) Move the cursor to the item corresponding to the H/W you want to reverse and touch the RTN button to switch to the data input mode.

3) The "H/W reverse" function reverses the actual operation signal, but does not change the display indicators.

2. Answer the questions in your own words.

1) What's the H/W reverse procedures?

2) Describe the stick calibration steps briefly.

3) If your friend want to fly quadcopter in American hand, please describe the setting procedures.

Mission 3   Model Select

【Objective】 Learn to add a new model, select the desired model, delete the model in the model list and change model name.

【Analysis】 This function is used to load the settings of the desired model into the T14SG's memory.

The settings may be selected from either the transmitter's internal memory or an SD card (see Fig. 2 - 3 - 1). Remember that up to 30 model memories are available in the transmitter.

The name of the model stored in the transmitter and the SD card may be changed. This can be very useful to tell different models' settings apart. Each model name can be as long as 10 characters, and the model name always appears in the display screen.

The copy function is used to copy parameters, settings, etc. from one model data into a second memory. It may be used for getting a head-start on setting up models with almost the same settings. Also, this function may be used to make a backup copy of a model setup before any changes are made.

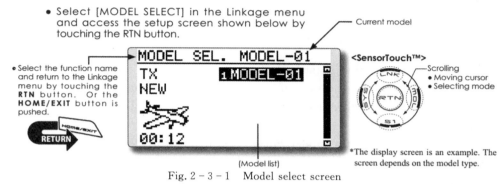

Fig. 2 – 3 – 1   Model select screen

【Knowledge preparation】 UAV classification.

2.3.1   Model Selection

(1)Move the cursor to the save destination display and touch the RTN button to switch to the data input mode.

(2)Select the save destination by scrolling the touch sensor and touch the RTN button.

(3)After moving the cursor to the desired model in the model list, touch the RTN button.

(4)Move to [SELECT].

(5)Touch the RTN button. A confirmation message is displayed (see Fig. 2 – 3 – 2). Touch the RTN button for one second and selection is complete.

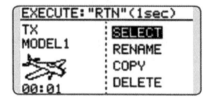

Fig. 2 – 3 – 2   Model selection confirmation

2.3.2   Model Addition

A new model can be added to the transmitter memory. It cannot be added to the SD card.

(1)Move the cursor to [NEW].

(2)Touch the RTN button. A confirmation message appears (see Fig. 2 – 3 – 3). Touch the RTN button for one second.

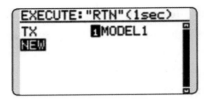

Fig. 2 - 3 - 3　Model addition confirmation

2.3.3　Model Deletion

The model stored in the transmitter memory or a SD card can be deleted.

(1)Move the cursor to the save destination display and touch the RTN button to switch to the data input mode. Select the save destination by scrolling the touch sensor and touch the RTN button.

(2)Move the cursor to the model you want to delete in the model list and then touch the RTN button.

(3)Move the cursor to [DELETE] (see Fig. 2 - 3 - 4).

(4)Touch the RTN button. When a confirmation message is displayed and the RTN button is touched for one second, the model is deleted.

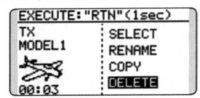

Fig. 2 - 3 - 4　Model deletion confirmation

2.3.4　Model Name Change

The model name of the model stored in the transmitter memory or a SD card can be changed.

(1)If changing the location, move the cursor to the save destination display and touch the RTN button to switch to the data input mode.

Select the save destination by scrolling the touch sensor and touch the RTN button.

(2)Move the cursor to the model you want to change in the model list and then touch the RTN button (see Fig. 2 - 3 - 5).

(3)Move to [RENAME].

(4)Touch the RTN button.

User name (candidate)

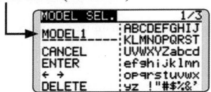

Fig. 2 - 3 - 5　Model name changing screen

(5)Change the model name as described below:

Moving cursor in the user name (candidate). Select [←] or [→], and touch the RTN button.

When [DELETE] is selected and the RTN button is touched, the character immediately after the cursor is deleted.

When a character is selected from the character list and the RTN button is touched, that character is added at the position immediately after the cursor.

After the desired information has been input, select [ENTER] and touch the RTN button.

After the desired information has been input, select [ENTER] and touch the RTN button. (To terminate input and return to the original state, select [CANCEL] and touch the RTN button.)

The model type setup screen and frequency setup screen are automatically displayed. Confirm or change the model type and SYSTEM mode.

Transmission stops and then starts in the new model.

The added model appears in the model list of the model select setup screen.

Link is required when a new model is made from a model selection.

Words & Phrases

confirmation [ˌkɒnfə'meɪʃn] n. 确认

terminate ['tɜːmɪneɪt] vt. 停止,结束

head-start [ˌhed 'staːt] n. 在某事开始时获得的优势,先机

Exercises and Thinking

1. Translate the following sentences.

1)Move the cursor to the save destination display and touch the RTN button to switch to the data input mode.

2)When a confirmation message is displayed and the RTN button is touched for one second, the model is deleted.

3)Link is required when a new model is made from a model selection.

2. Answer the questions in your own words.

1)What's the function of "MODEL SELECT"?

2)Describe the steps of "Model Deletion".

3)Describe the steps of "Model Name Change".

3. Read the following passages and answer questions.

Model Copy

A copy can be made of the model stored in the transmitter memory or a SD card.

1)If changing the location:

Move the cursor to the save destination display ("TX" or "CARD") and touch the RTN button to switch to the data input mode.

Select the save destination by scrolling the touch sensor and touch the RTN button.

[TX]:Transmitter memory.

[CARD]:SD card.

2) Select the model you want to copy in the model list and then touch the RTN button.

3) Move to [COPY].

4) Touch the RTN button. The copy screen appears as Fig. 2 - 3 - 6.

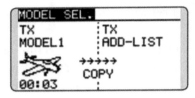

Fig. 2 - 3 - 6   The copy screen

5) If replacing the model stored in the transmitter memory:

Move to [ADD-LIST] and touch the RTN button to switch to the data input mode.

Select the destination model by scrolling the touch sensor and touch the RTN button.

[ADD-LIST]:adding the model to the list.

[(model name)]:replacing the model.

If changing the location:

Move the cursor to the copy destination display ("TX" or "CARD") and touch the RTN button to switch to the data input mode.

Select the save destination by scrolling the touch sensor and touch the RTN button.

6) Move to [COPY].

7) Touch the RTN button. When a confirmation message is displayed and the RTN button is touched for one second, the model data is copied.

Questions:

1) Describe the basic steps of "Model Copy".

2) What should you do if replacing the model stored in the transmitter memory?

Mission 4   Model Type

【Objective】 Select the model type for your UAV.

【Analysis】 This function selects the model type from among airplane, helicopter, and glider (see Fig. 2 - 4 - 1).

Six swash types are available for helicopters. Six types of main wings and three types of tail wings are available for airplanes and gliders. Functions and mixing functions necessary for each model type are set in advance at the factory.

When the Model Type selection command is accessed, all of the data in the active memory is cleared. Be sure that you don't mind losing this data, or back it up to another memory using the copying functions.

When changing the helicopter swash type (see Fig. 2 – 4 – 2) within the following groups, you can leave the settings other than the SWASH function. However, this is initialized when you change the swash type to the other swash type group.

- Select [MODEL TYPE] in the Linkage menu and access the setup screen shown below by touching the RTN button.

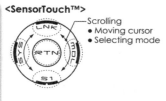

(The display screen is an example. The screen depends on the model type.)

Fig. 2 – 4 – 1　Model type screen

Swash type group A:
H-1, H-3, HR3, and HE3
Swash type group B:
H-4, H-4X

Fig. 2 – 4 – 2　Swash type group

【Knowledge preparation】 Flight principle of helicopter; the structure of airplane.

The procedures of model type selection are depicted in detail.

(1) Move the cursor to the item you want to change and touch the RTN button to switch to the data input mode (see Fig. 2 – 4 – 3).

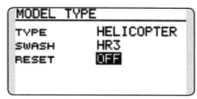

Fig. 2 – 4 – 3　Change the model type

(2) Select the desired type by scrolling the touch sensor and touch the RTN button. A confirmation message appears. Touch the RTN button for one second.

(3) Move to [YES] and Touch the RTN button for one second.

"TYPE": Model type.

"WING" (airplane/glider): Wing type (see Fig. 2 - 4 - 4).

"TAIL" (airplane/glider): Tail type.

"SWASH" (helicopter): Swash type.

If resetting the data when changing the helicopter swash type (see Fig. 2 - 4 - 5):

(1)Move the cursor to [OFF] and touch the RTN button to switch to the data input mode.

(2)Select [ON] by scrolling the touch sensor and touch the RTN button. A confirmation message appears. Touch the RTN button.

(3)Activate the swash type setting. The swash setting parameters are reset.

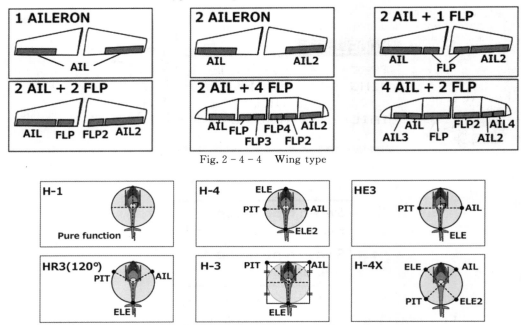

Fig. 2 - 4 - 4　Wing type

Fig. 2 - 4 - 5　Swash type for helicopter

The Model Type function automatically selects the appropriate output channels, control functions, and mixing functions for the chosen model type. Before you select model type, you should know the UAV you are flying well and be aware of the structure of UAV.

Words & Phrases

helicopter ['helɪkɒptə(r)] *n.* 直升机

glider ['glaɪdə(r)] *n.* 滑翔机

swash type [swɒʃ] [taɪp] 十字盘

scroll [skrəʊl] *v.* 滚动

Exercises and Thinking

1. Translate the following sentences.

1) This function selects the model type from among airplane, helicopter, and glider.

2) Functions and mixing functions necessary for each model type are set in advance at the factory.

3) The model type function automatically selects the appropriate output channels, control functions, and mixing functions for the chosen model type.

2. Answer the questions in your own words.

1) What's the function of "MODEL TYPE"?

2) Describe the procedures of Model Type selection briefly.

Mission 5　Reverse and End Point

【Objective】 Use to reverse the throw direction; set the travel and limit point of each servo.

【Analysis】 Servo Reverse changes the direction of an individual servo's response to a control input. For CCPM (Collective-Cyclic Pitch Mixing) helicopters, be sure to be familiar with Swash AFR (Adjustable Function Rate) before reversing any servos. With CCPM helicopters, always complete your servo reversing prior to any other programming. If you use pre-built Airplane/Sailplane functions that control multiple servos, it may be confusing to tell whether the servo needs to be reversed or a setting in the function needs to be reversed. See the instructions for each specialized function for further details. Always check servo direction prior to every flight as an additional precaution to confirm proper model memory, hook ups, and radio function.

The End Point function adjusts the left and right servo throws, generates differential throws, and will correct improper linkage settings. The travel rate can be varied from 0% to 140% in each direction on channels 1 to 12. Also, the limit point where servo throw stops may be varied from 0% to 155%.

【Knowledge preparation】 The structure of UAV; the flight control of UAV.

2.5.1　Reverse

The following are the basic servo reversing procedures (see Fig. 2 − 5 − 1).

(1) Move the cursor to the channel you want to reverse and touch the RTN button to switch to the data input mode.

(2) Select the direction by scrolling the touch sensor. A confirmation message appears.

(3) Touch the RTN button to change the direction (To terminate input and return

to the original state, touch the S1 button).

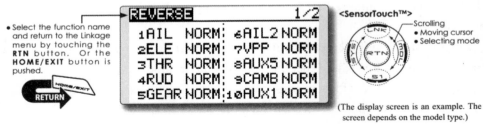

Fig. 2 – 5 – 1 Servo reversing procedures

Tips: Upon setup completion of a new model, check whether or not each servo is connected to the correct channel.

Next, determine whether you need to reverse any channels by moving each stick and/or other control inputs.

Repeat the operation above for each channel that must be reversed.

2.5.2 End Point

1. Servo travel adjustment

(1)Move the cursor to the travel icon of the channel you want to adjust and touch the RTN button to switch to the data input mode.

(2)Adjust the rate by scrolling the touch sensor.

Initial value: 100%.

Adjustment range: 0% – 140%.

When the RTN button is touched for one second, the rate is reset to the initial value. Touch the RTN button to end adjustment and return to the cursor mode.

(3)Repeat this procedure for each rate.

2. Limit point adjustment

(1)Move the cursor to the limit point icon of the channel you want to adjust and touch the RTN button to switch to the data input mode.

(2)Adjust the limit point by scrolling the touch sensor.

Initial value: 135%.

Adjustment range: 0% – 155%.

When the RTN button is touched for one second, the limit point is reset to the initial value. Touch the RTN button to end adjustment and return to the cursor mode.

Repeat this procedure for each limit point.

End Point display screen see Fig. 2 – 5 – 2.

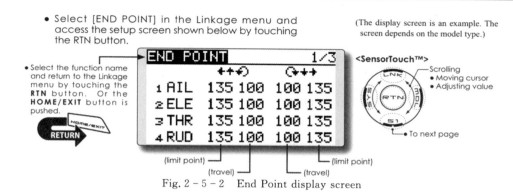

- Select [END POINT] in the Linkage menu and access the setup screen shown below by touching the RTN button.

(The display screen is an example. The screen depends on the model type.)

• Select the function name and return to the Linkage menu by touching the **RTN** button. Or the **HOME/EXIT** button is pushed.

<SensorTouch™>
Scrolling
• Moving cursor
• Adjusting value

• To next page

(limit point)　(travel)　(travel)　(limit point)

Fig. 2 - 5 - 2　End Point display screen

When you want to reverse the channel, you can use the "Reverse" function. The "End Point" function is suitable for adjusting the servo throws which can change the deflection angle of control surfaces.

Words & Phrases

CCPM＝Collective-Cyclic Pitch Mixing [kə'lektɪv] ['saɪklɪk] [pɪtʃ] ['mɪksɪŋ] 螺距混控系统

AFR＝Adjustable Function Rate [ə'dʒʌstəbl] ['fʌŋkʃn] [reɪt] 可调功能比率

hook up ['hʊk ʌp] 连接

throw [θrəʊ] n. 行程

Exercises and Thinking

1. Translate the following sentences.

1) Servo reverse changes the direction of an individual servo's response to a control input.

2) If you use pre-built Airplane/Sailplane functions that control multiple servos, it may be confusing to tell whether the servo needs to be reversed or a setting in the function needs to be reversed.

3) Move the cursor to the travel icon of the channel you want to adjust and touch the RTN button to switch to the data input mode.

2. Answer the questions in your own words.

1) What's the function of "Reverse"?

2) What's the function of "End Point"?

3) Describe the basic operations of "Reverse" briefly.

Mission 6　Failsafe

【Objective】 Set up positions that the servos move to in the case of radio interference.

【Analysis】 You may set either of two positions for each channel: Hold, where the servo maintains its last commanded position, or Failsafe, where each servo moves to a predetermined position (see Fig. 2 - 6 - 1). You may choose either mode for each channel.

The T14SG system also provides us with an advanced battery monitoring function that warns you when the receiver battery has only a little power remaining. In this case, each servo is moved to the defined failsafe position. The battery failsafe may be released by operating a predefined control on the transmitter (default is throttle), do not continue to fly, land as soon as possible. Remember, if the predefined control suddenly moves to a position you did not command, land at once and check your receiver battery.

Defines servo position when signals are lost and when receiver battery voltage becomes low.

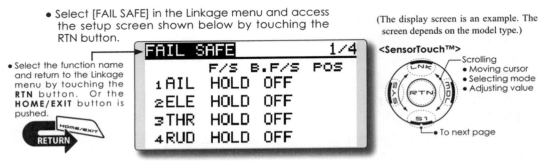

Fig. 2 - 6 - 1   Failsafe procedures

【Knowledge preparation】 The common fault of UAV.

2.6.1   Fail Safe Setting Procedure

(1)Move the cursor to the "F/S" item of the channel you want to set and touch the RTN button to switch to the data input mode.

(2)Select the F/S mode by scrolling the touch sensor. A confirmation message appears, and the display blinks.

(3)Touch the RTN button. The channel switches to the F/S mode.

(4)Move the cursor to the "POS" item. Hold the corresponding stick, knob, slider, etc. in the position you want the servo to move to when the failsafe function is activated and touch the RTN button for one second.

• The set position is displayed in percentage.

• If you want to return that channel to the hold mode, move the cursor to the "F/S" item and touch the RTN button to switch to the data input mode. Select the F/S mode by scrolling the touch sensor. A confirmation message appears and then change the mode by touching the RTN button.

2.6.2   Battery Failsafe Setting Procedure

Battery failsafe can be set for each channel by the same method as the failsafe setting procedure. Select and set the "B. F/S" item.

[ON]: Battery fail safe function ON.

[OFF]: Battery fail safe function OFF.

2. 6. 3　Battery Failsafe Release Switch Setting

This function temporarily releases the battery failsafe function, so the fuselage can recover after the battery failsafe function was activated by a drop in the receiver battery voltage. This setting selects the switch which releases the battery fail safe function.

(1) Move the cursor to the [RELEASE B. F/S] item in the setup screen.

(2) Touch the RTN button.

For safety, always set the failsafe functions.

• Remember to set the throttle channel fail safe function so that the servo moves to the maximum slow side for airplanes and to the slow side from the hovering position for helicopters. Crashing of the model at full high when normal radio waves cannot be received due to interference, etc. , is very dangerous.

• If the battery failsafe is reset by the throttle stick, it may be mistaken for an engine malfunction and will be reset at throttle slow and the model will continue to fly. If you have any doubts, immediately land.

Words & Phrases

failsafe ['feɪlˌseɪf] *n.* 失控保护

blink [blɪŋk] *v.* 闪烁

slider ['slaɪdə(r)] *n.* 滑块

fuselage ['fjuːzəlɑːʒ] *n.* 机身

malfunction [ˌmælˈfʌŋkʃn] *n.* 故障

Exercises and Thinking

1. Translate the following sentences.

1) The T14SG system also provides us with an advanced battery monitoring function that warns you when the receiver battery has only a little power remaining.

2) The battery failsafe may be released by operating a predefined control on the transmitter (default is throttle), do not continue to fly, land as soon as possible.

3) If the battery failsafe is reset by the throttle stick, it may be mistaken for an engine malfunction and will be reset at throttle slow and the model will continue to fly.

2. Answer the questions in your own words.

1) Introduce the failsafe setting procedure to your classmates.

2) Why do we need to set the failsafe functions?

3) Depict the battery failsafe setting procedure in brief.

Mission 7　Throttle Cut

【Objective】Choose the location and direction of a switch to stop the engine.

【Analysis】Throttle Cut provides an easy way to stop the engine. Generally speaking, modelers will do so by flipping a switch with the throttle stick at idle. The action is not functional at high throttle to avoid accidental dead stick landings. The switch's location and direction must be chosen, as it defaults to NULL.

If the Throttle Cut switch is activated, or on, this status will continue even if the condition is changed to an inhibited setting.

If the condition is inhibited (INH), the Throttle Cut is off if the SW is in the off position and the throttle stick is low.

【Knowledge preparation】The flight dynamic of UAV.

Fig. 2 – 7 – 1 shows the basic procedures of THR CUT setting. The following are Throttle Cut setting procedures.

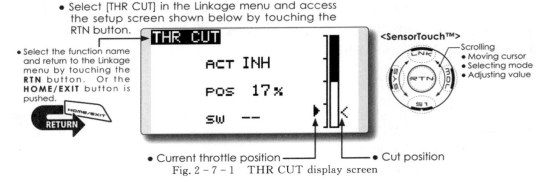

Fig. 2 – 7 – 1　THR CUT display screen

(1)Activate the function.

Move the cursor to the [ACT] item and touch the RTN button to switch to the data input mode. Select the ACT mode by scrolling the touch sensor. Touch the RTN button to activate the function and return to the cursor mode.

(2)Switch selection.

Move the cursor to the [SW] item and access the switch setup screen by touching the RTN button and select the switch and ON direction.

(3)Throttle cut position setting.

Move the cursor to the [POS] item and touch the RTN button to switch to the data input mode. Adjust the servo operation position at Throttle Cut operation by scrolling the touch sensor.

Since conditions are not offered when an Airplane is selected, the Throttle Cut options will vary from the options.

Individually adjust the Throttle Cut activation setting for each condition.

The Throttle Cut POS and SW settings are utilized for all conditions.

Words & Phrases

flip〔flɪp〕*v.* 快速翻转

accidental〔ˌæksɪ'dentl〕*adj.* 意外的

dead stick landing〔ded〕〔stɪk〕〔'lændɪŋ〕*n.* 滑翔着陆（引擎故障时的飞机降落）

inhibited〔ɪn'hɪbɪtɪd〕*adj.* 禁止的，抑制的

Exercises and Thinking

1. Translate the following sentences.

1) Generally speaking, modelers will do so by flipping a switch with the throttle stick at idle.

2) If the Throttle Cut switch is activated, or on, this status will continue even if the condition is changed to an inhibited setting.

3) Since conditions are not offered when an Airplane is selected, the Throttle Cut options will vary from the options.

2. Answer the questions in your own words.

1) Why do we need to set the "Throttle Cut" function?

2) Depict the basic setting procedures of "Throttle Cut".

Mission 8　Dual Rate and Throttle Curve

【Objective】Adjust the curve of aileron, elevator, rudder and throttle.

【Analysis】Dual rate function is used to adjust the amount of throw and the operational curve of the stick functions (aileron, elevator and rudder) for each flight condition or up to 5 rates for each function. For airplane type, it is also possible to adjust the operational curve of the throttle function. This is normally used after the End Point programming has been completed to define the maximum throw. When mixing is applied from one channel to another channel, both channels can be adjusted at the same time by adjusting the operation rate through the dual rate function.

Throttle curve function adjusts the throttle curve for optimum engine speed from throttle stick input. When throttle curve is set to ON when there is no throttle function; this curve acts as the motor function.

【Knowledge preparation】Flight dynamics of UAV.

2.8.1　Dual Rate

The following are Dual Rate setting procedures.

(1) Function selection.

Move the cursor to the function selection item and touch the RTN button to switch to the data input mode. Select the function you want to adjust by scrolling the touch sensor. Touch the RTN button to the cursor mode.

(2)Switch selection.

Move the cursor to the circuit # item and access the switch setup screen by touching the RTN button. Select the switch activation method and the activation position (if applicable).

(3)Left/right (up/down) rate adjustment.

Move the cursor to the rate item you want to adjust and touch the RTN button to switch to the data input mode (see Fig. 2-8-1). Adjust the rate by scrolling the touch sensor. Touch the RTN button to end the adjustment and return to the cursor mode.

Repeat this procedure for additional rate and other functions as desired.

(4)Operation curve (EXP curve) adjustment.

Move the cursor to the EXP item you want to adjust and touch the RTN button to switch to the data input mode. Adjust the rate by scrolling the touch sensor.

Initial value:0%.

Adjustment range:-100%-+100%.

Touch the RTN button to end adjustment and return to the cursor mode.

Repeat this procedure for all other rates and functions as desired.

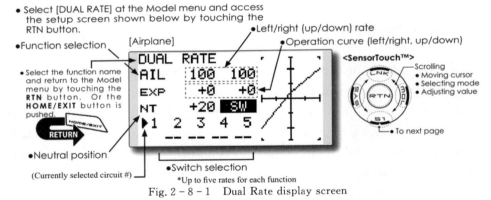

Fig. 2-8-1 Dual Rate display screen

2.8.2 Throttle Curve

The following are setting method.

1.Activate the function

Move the cursor to the [ACT] item and touch the RTN button to switch to the data input mode.

Select the ON mode by scrolling the touch sensor.

Touch the RTN button to activate the function and return to the cursor mode.

2.5-point curve setting

For curve rate setting, we should do the following steps.

(1)Move the cursor to the curve rate setting item you want to adjust and touch the RTN button to switch to the data input mode (see Fig. 2 - 8 - 2). Adjust the rate by scrolling the touch sensor. Touch the RTN button to end the adjustment and return to the cursor mode.

(2)Repeat this procedure for each point.

For moving curve point, the steps below are necessary.

(1)Move the cursor to the curve point setting item you want to adjust and touch the RTN button to switch to the data input mode. Adjust the curve point by scrolling the touch sensor.

Initial value: P1:(0%), P2:25%, P3:50%, P4:75%, P5:(100%).

Adjustment range: Up to 2.5% in front of the adjoining point.

Touch the RTN button to end the adjustment and return to the cursor mode.

(2)Repeat this procedure for each point.

When you need to delete/return curve point, you should move the cursor to the curve point setting item you want to delete/return and touch the RTN button for one second.

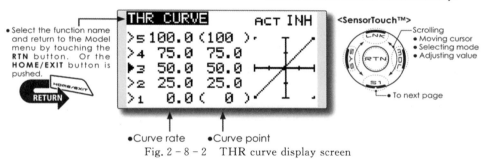

Fig. 2 - 8 - 2　THR curve display screen

Using the EXP curve is effective to smooth or soften the control inputs around center to avoid over-controlling the model. This is often used for the ailerons, elevator and rudder and may be used with the throttle in the case of an airplane selection to smooth the engine controls as well.

The throttle curve adjustment is available for the corresponding model type: airplane, glider and general.

Words & Phrases

curve [kɜːv] n. 曲线

dual rate ['djuːəl] [reɪt] n. 双比率开关

adjoining [əˈdʒɔɪnɪŋ] *adj.* 相邻的，邻接的

Exercises and Thinking

1. Translate the following sentences.

1) When mixing is applied from one channel to another channel，both channels can be adjusted at the same time by adjusting the operation rate through the dual rate function.

2) Throttle curve function adjusts the throttle curve for optimum engine speed from throttle stick input.

3) Using the EXP curve is effective to smooth or soften the control inputs around center to avoid over-controlling the model.

4) Always check servo direction prior to every flight as an additional precaution to confirm proper model memory，hook ups，and radio function.

5) The T14SG system also provides us with an advanced battery monitoring function that warns you when the receiver battery has only a little power remaining.

6) Using the EXP curve is effective to smooth or soften the control inputs around center to avoid over-controlling the model.

2. Answer the questions in your own words.

1) Why do we need to set the "Dual Rate" function?

2) What's the function of "Throttle Curve"?

3) According to what you have learned in this article，which operation of radio control system is most difficult? Why?

4) If you have a Cessna model airplane and a Futaba 14SG，and you want to fly this model airplane，try to descript the basic operations of radio control system.

5) Please give a brief introduction of failsafe setting procedures.

【Project Evaluation】

Work activities		Weightage	Scores
Reading and translation		10	
Link the receiver with the transmitter		6	
H/W setting	Reverse the operation direction	3	
	Change stick mode	4	
	Calibrate stick	4	
Model Select	Select a model	2	
	Add a new model	3	
	Delete a model	2	
	Change model name	2	
Select the model type		3	
Reverse the servo response direction		5	
Set the Fail Safe function for each channel		9	

Continued

Work activities	Weightage	Scores
Use end point function to adjust servo throws	10	
Set throttle cut function for a switch	9	
Change the curve of the stick functions	10	
Change the throttle curve	10	
Attitude and attendance	8	

Grade: □Excellence   □Good marks   □Medium level   □Pass

【Project Conclusions】Upon completion of this project, you shall be able to:

1)Select radio controller for drone;

2)Establish communication link between transmitter and receiver;

3)Add a new model and select the appropriate model type for your drone;

4)Set the fail safe function and throttle cut function for your fixed wing drone;

5)Adjust the left and right servo throws;

6)Change the aileron, elevator, rudder and throttle curve for your drone to make it have the best performance.

Project 3　UAV Assembly and Commissioning

【Description】 This project will guide you to assemble your own drone and conduct drone commissioning making the drone fly safely and carry out tasks, you will need equipments, tools, material for UAV assembly and commissioning.

【Analysis】 You should preview the key words in advance and grasp the meaning of them, then listen to the teacher carefully for the operating procedures, and practice repeatedly in groups.

【Related knowledge and skills】 The structure and systems of UAV; circuit and electronic technology; soldering; operating the GCS.

Mission 1　Quadcopter Assembly and Commissioning

【Objective】 Build a quadcopter for myself.

【Analysis】 Build our own quadcopter can bring us satisfaction. As students of UAV speciality, we cannot satisfactory with just flying ARF multicopter. We should also master the maintenance skill, which can be acquired by doing this mission. Rather than purchase a kit or follow a set of online instructions, we spend a lot of time researching quadcopters, and eventually can put together a thorough tutorial myself. In the construction process, we are encouraged to discuss frame fabrication and component placement as well as how to program the APM for the copter's first flight in groups which can also cultivate the spirit of teamwork.

【Knowledge preparation】 UAV structure and systems; electromagnetics; circuit; soldering.

3.1.1　What We Need?

The DJI F330 FlameWheel is inexpensive, sturdy and small. It can be flown indoors or out. Even though it is compact, the F330 is substantial and can easily carry a battery that will permit 15-minute flights. The F330 can also carry a GoPro type camera or FPV system if desired.

The ARF (Almost Ready to Fly) kit is a super bargain and includes motors, ESCs (Electronic Speed Controls) and Propellers. The DJI Motors and ESCs provided in the ARF kit are of high quality, very durable and long lasting. And the frame provided is extremely rugged and all replacement parts are readily available and inexpensive.

The DJI F450 Flamewheel is an excellent larger alternative to the F330. The F450 FlameWheel ARF kit has a larger frame, 30 Amp ESCs and 10 in(1in＝2.54 cm) props for only ￥10.00 more than the F330.

The F450 is well suited to carrying a GoPro camera, brushless gimbal, landing gear and a larger battery for longer flights.

The F330 and F450 FlameWheel are excellent platforms for the PX4 and Pixhawk Flight controllers as well as the APM.

It's strongly recommended to use the F330 and F450 ARF kits, because they are reliable, rugged and inexpensive and are suitable for many uses.

The quality is excellent, the components provide long trouble-free service, and replacement parts are cheap and available.

We cannot accomplish an entire multicopter only with equipment introduced above. There are basic components (see Fig. 3 - 1 - 1) required in addition to the Flamewheel ARF kit. Let's see below:

• A Flight Control Board (APM 2.6, PX4FMU ＋ PX4IO or Pixhawk) with UBlox GPS module and power supply module.

• RC equipment: RC transmitter and a 5＋ channel radio receiver (PPM-SUM compatible for PX4 or Pixhawk).

• A 3 cell 1,800mAh to 2,650mAh LiPo battery and an battery Charger.

• Miscellaneous: female to female 3 wire receiver/servo cables, a battery strap, tie wraps and some assorted hardware.

You will also need an adjustable soldering iron, rosin cored solder, blue thread locker and some small Metric Allen wrenches.

The whole tools/materials/equipments used to assemble and commision a quadcopter are listed in Table 3 - 1 - 1.

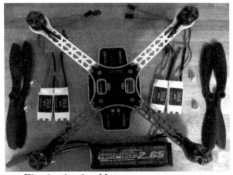

Fig. 3 - 1 - 1   Necessary components

Table 3 − 1 − 1　Tool/equipment/material

Type	Name	TS	SI	Num
Tool	Soldering iron	50 W		1
	Insulating varnish stripper	JRF-CSP-501		1
	Tweezer			1
	Diagonal plier	3.5 in		1
	Needle-nose plier			1
	Solder sucker			1
	Allen wrench	2 mm		1
	Allen wrench	2.5 mm		1
	Silicone lead (red)	14♯AWG	m	0.2
	Silicone lead (black)	14♯AWG	m	0.2
	Solder paste			1
	Solder wire			1
Material	Upper center board			1
	Power distribution board			1
	Motor	2212 KV1400		4
	Frame arm			4
	ESC	30 A		4
	WFLY RC equipment	2.4 GHz		1
	APM Flight Controller Board	Rev. 2.8		1
	UBEC	5 V,3 A		1
	Frame arm screw	M2.5×6		24
	Motor screw	M3×8		16
	Velcro		m	0.2
	Velcro strap	20 cm		10
	Battery	3 s 2,200 mAh		1
	Dupont line			5
	ATG propeller	10×4.7		4
	Heat shrink tube	4 mm	m	0.2
Equipment	Computer	Win7		1
	RC	2.4 GHz		1
	Battery	1 s 2,200 mAh		1
	USB Data Cable	2.0	m	1

3.1.2   ESC Installation and Soldering the Power Distribution Board

1. Install the ESCs with tie wraps

Thread the Tie Wraps from front to back through the frame members and between the motor bullet connector sockets.

Install the ESC power leads through the slot under the bottom of the frame member (see Fig. 3 – 1 – 2).

Then trim and solder them to the power distribution board.

This is a little harder than going around the outside of the frame but is neater and provides a little more battery clearance.

The above illustration shows a power distribution "Y" connector which functions the same as the FlameWheel's built in power distribution board.

2. Soldering the power distribution board

Soldering the ESCs to the Power Distribution Board (bottom frame plate) does require a bit of creativity when doing this. Because we need to adjust the cable direction (see Fig. 3 – 1 – 3) to make the subsequent frame assembly more convenient and fit things more closely.

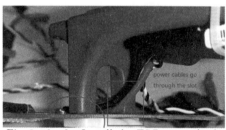

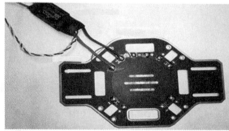

Fig. 3 – 1 – 2   Install the ESC power leads　　　Fig. 3 – 1 – 3   Soldering the ESC to PDB

When you solder the ESC and Battery power leads to the power distribution board. Remember to flux and preheat both the pads and the wire in order to get a properly tinned (wet) solder joint.

The board can soak up quite a bit of heat, so make sure your soldering iron sufficiently hot.

The DJI ARF FlameWheel kits have Opto ESCs which do not include a BEC so you will need an external power supply.

The APM, PX4 and Pixhawk are now available with a power supply you can use, otherwise you will need to buy a BEC. And there are many companies that make an excellent and very reliable 10/5 Amp BEC.

Use a switching BEC in any case, they are more reliable and a lot more efficient than linear ones.

You do not need to and in fact can't balance the DJI ESCs as shown elsewhere in the Copter Wiki.

DJI ESCs are digital, optically isolated, are completely pre-balanced at the factory and work fine out of the box.

3.1.3　Frame and Motor Assembly

(1)Preparation.

1)Check the availability of tools, equipment and material.

2)Prepare tools, equipment, material.

3)Choose effective technical documents.

4)Check if the types and number of material are complete.

5)Be aware of the correct assembly procedures.

(2)Wear esd wrist strap.

(3)Study the assembly diagram (see Fig. 3 - 1 - 4) carefully, make clear the connections between different components.

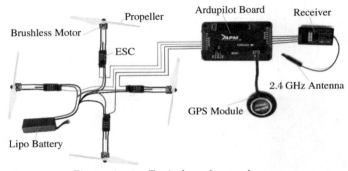

Fig. 3 - 1 - 4　Typical quadcopter layout

(4)Use blue or purple removable thread locker very sparingly on the threads of the Allen screws when you assemble the frame (see Fig. 3 - 1 - 5). Try to keep excess thread locker off the plastic. This is actually not achievable, but do the best you can.

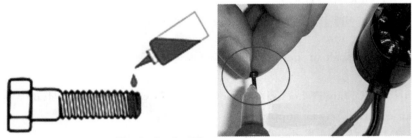

Fig. 3 - 1 - 5　Thread locker sealing

(5)Motor and frame screws have a shoulder on them and you need to tighten them all sufficiently so that the shoulder fully seats. A good method is to install all screws (with Locktite) loosely then tighten all the screws on each frame arm one at a time. Do this in one sitting.

(6)When installing the motors, pre-bend the motor leads upward where they exit the motor bottom so that the motor can sit flat. Have the leads exit towards the center of the copter. Motor bottom screws are off set and it can only be installed in 2 ways, toward the center is correct. The motor screws are short and it is tricky to start the first screw on the motor bottom. Ensure that the motor leads are not pushing the motor up. Use blue Locktite on each screw and install all 4 screws for a motor loosely and then tighten them.

(7)Thread the motor leads down through the first 2 large frame member holes. Or you can run them around the frame member if you prefer.

3.1.4   Vibration Damping

The Flight controllers require special attention be paid to removing as much vibration from the flight control board as possible.

This F330 FlameWheel has a special anti-vibration suspension mounting of the APM flight controller. It uses a folded over 1/16 in O-ring at each corner of the flight control board around a screw sticking out of a small standoff. If you choose to use this method leave 1/10 in to 1/8 in clearance from each board corner for proper short coupled spacing. And the F330 usually requires minor surgery with a Dremel tool (see Fig. 3 – 1 – 6) on each frame arm to achieve proper hole spacing in the top plate.

There are the basic procedures that reduce vibration installation of the flight control board.

(1)Simply put a 3/4-inch square of adhesive backed Kyosho Zeal Gel (see Fig. 3 – 1 – 7) under each corner of the flight controller. Kyosho Zeal Gel is easy to install and fully satisfactory alternative to the O-ring suspension method. The Gel is a material that possess excellent vibration dampening characteristics, ideal for mounting gyroscopes, receivers, or other electronics in vibration prone areas. Its excellent vibration dampening characteristic can minimize impact induced damages on electronics.

Fig. 3 – 1 – 6   A dremel tool   Fig. 3 – 1 – 7   Kyosho Zeal Gel

(2)Secure the flight controller with a Velcro strap over 1/2 in of soft foam in light tension. (Do NOT over tighten!)

Either the Kyosho Zeal Gel or the O-ring suspension method (see Fig. 3 – 1 – 8) will allow you to limit vibration to about 1/10 G which is fine.

More anti-vibration techniques, methods and "tuning" are covered in greater detail online, you can search on the internet and share what you found in class.

Fig. 3 − 1 − 8   Anti-vibration installation of the flight control board

3.1.5   Connect ESCs and Motors

1. Connect motor PWM signal outputs

When connecting the ESCs directly to autopilot board, connect the power (＋), ground (−), and signal(s) wires for each ESC to the controller main output pins by motor number. Find your frame type below to determine the assigned order of the motors.

Make sure you connect the ESC connector (see Fig. 3 − 1 − 9) in the right way. Signal goes on the top of the rail (white or orange color wire) and ground at the bottom (black or brown color wire).

Fig. 3 − 1 − 9   The ESC connector

2. Connect motor PWM signal outputs (APM2)

There are two methods of connecting the motor outputs: connect the electronic speed controllers (ESCs) to autopilot controller board directly or use a power distribution board (PDB).

When using a PDB, connect the power (＋), ground (−), and signal(s) wires for each ESC to the PDB according to motor number (see Fig. 3 − 1 − 10). Determine the assigned order of the motors according to your frame type. Then connect the signal wires from the PDB to the main output signal pins on the flight controller board (ensuring that the motor order numbers match the main output pin numbers on the controller). If you are using a power module, it is optional to connect the power and ground wires from the PDB to the flight controller board. If you would like to use these cables in addition to or instead of the power module or as a common point for low current servos, connect the ground (−) wire to a main output ground (−) pin and the power (＋) wire to a main output power (＋) pin.

Fig. 3 – 1 – 10   APM output pins

3. Understand motor order diagrams

The sections below show motor order for each frame type (the numbers indicates the connected autopilot output pin)and the propeller direction (clockwise (CW) motors take pusher propellers, counterclockwise motors (CCW) take puller propellers).

Use the diagram for your frame type (see Fig. 3 – 1 – 11), and wire the motors as shown.

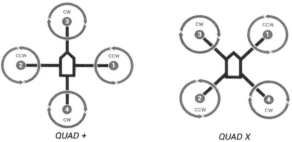

Fig. 3 – 1 – 11   Frame type

4. Choose the appropriate propellers

The stock 8 in DJI props that are supplied with the F330 ARF kit work fine. However, the F330 FlameWheel can definitely benefit from a 9 in prop (10 in will not fit on the F330). GemFan 9 in multirotor "Carbon Filled" props are only $2. 00 each and are superior to the stock 8 in props. These GemFans are almost impossible to break and the 9 in ones are considerably more efficient than the stock 8 in props. 9 in GemFan propellers are not made to fit the oval hubs of the DJI Motors. But you can carefully bore them out to about. 31 in with an ordinary drill and they will fit perfectly. On F450 Flamewheel, you can use 11 in GemFan propellers bored out the same way. Generally, a larger prop diameter is better so long as you don't exceed motor or ESC maximums or cause overheating.

Make sure the propellers are right side up (printing on top), this won't fly worth a damn if they are upside down.

5. Attach propellers

Find your frame in the motor order diagrams above. Clockwise motors are shown in green, marked CW, and take pusher propellers. Counterclockwise motors are shown in blue, marked CCW, and take puller propellers (see Fig. 3 – 1 – 12). Use the diagram for your frame type and attach propellers to your vehicle as shown. For copters, attach propellers with the writing facing towards the sky. For more information on recognizing the different types of propellers, see the next section.

Fig. 3 - 1 - 12　CW and CCW

6. Recognizing clockwise and counterclockwise propellers

The diagrams above show two types of propellers: clockwise (called pushers) and counterclockwise (called pullers). Pusher propellers are often marked with a P. However, not all propellers are marked and both types are often available in either rotational direction. Therefore, it is most reliable to recognize the correct propeller type by its shape as shown below. As shown in Fig. 3 - 1 - 13, the propellers have the edge with the shallow consistent curve at the leading edge in direction of rotation and the more radical scalloped (and usually thinner edge) as the trailing edge. You can use these features to recognize propellers of the correct direction of rotation.

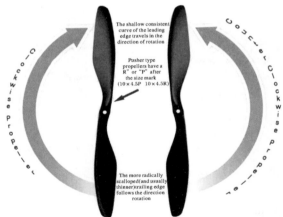

Fig. 3 - 1 - 13　Recognizing clockwise and counterclockwise propellers

7. Testing motor spin directions

If you have completed the Radio and ESC calibration, you can check that your motors are spinning in the correction direction:

(1)Make sure there are no propellers on your copter!

(2)Turn transmitter on and ensure the flight mode switch is set to Stabilize.

(3)Connect battery.

(4)Arm copter by holding the throttle down and rudder right for five seconds.

(5)If it fails to Arm with the throttle down and to the right and the motors will not spin, it has probably failed the Pre-arm Safety Check. Pre-arm Safety Check failure is also indicated by the red arming light double flashing and then repeating. If the Pre-arm check fails go to the Pre-arm Safety Check Page and correct the problem or disable the check before continuing.

(6)When you can Arm successfully, apply a small amount of throttle, and observe and note spin direction of each motor. They should match directions shown in the images above for the frame you've chosen.

(7)Reverse any motor spinning in the wrong direction.

3.1.6　Install Battery

Our construction method allows the battery to be inserted between the two frame plates. This is an optimal location regarding the quadcopter's center of gravity. The 2,650 mAh LiPo Turnigy Nano battery used here is as large as will comfortably fit, but it will permit 15-minute plus flights.

(1)Cover bottom frame power distribution solder joints with liquid electrical tape or silicone to prepare for the battery.

(2)Wrap the battery in bubble wrap and tape it in place to provide additional electrical isolation from the solder joints. This will also provide cushioning from vibration and sub-optimal "landings".

(3)Use a velcro battery strap through the frames bottom slots to retain the battery.

(4)Check if all assembly procedures are done.

(5)Sort and count tools and equipment.

(6)Clean the work site.

3.1.7　Quadcopter Commissioning

(1)Preparation.

1)Check if computer can work normally and network is smooth.

2)Read the MissionPlanner instruction manual in detail.

3)Check if the types and number of materials are complete.

4)Check the availability of tools, equipments and materials.

(2)Turn on the computer, download and install NetFramework 4.0. Install Mission Planner on the computer.

(3)Open MP and use usb data cable to connect flight controller to the computer. Set the Baud rate of serial port to 115,200 and then click "connect" button.

(4)Check if the plane type of flight controller is QUADX. If not, you must rewrite correct firmware to flight controller.

(5)Compass calibration.

The aim is to rotate the vehicle so that the colored trail hits each of the white dots. One way to do this is to hold the vehicle in the air and rotate it slowly so that each side (front, back, left, right, top and bottom) points down towards the earth for a few seconds in turn.

The calibration will automatically complete when it has data for all the positions. At this point, another window will pop up telling you that it is saving the newly calculated offsets. These are displayed on the main screen below each associated compass.

(6)Accelerometer calibration.

Under Initial Setup | Mandatory Hardware, select Accel Calibration from the left-side menu.

Click Calibrate Accel to start the calibration. Mission Planner will prompt you to place the vehicle each calibration position. Press any key to indicate that the autopilot is in position and then proceed to the next orientation.

The calibration positions are: level, on right side, left side, nose down, nose up and on its back.

Proceed through the required positions (it is not necessary to use the Click When Done button).

(7)RC transmitter flight mode configuration.

Turn on your RC transmitter. Connect the Pixhawk (or other flight controller) to the Mission Planner. Go to the Initial Setup | Mandatory Hardware | Flight Modes screen.

Use the drop-down on each line to select the flight mode for that switch position.

Ensure that at least one switch position is left assigned to STABILISE.

Optionally check the Simple Mode check-box for that switch position. If using AC3.1 or higher you can optionally set Super Simple mode. If both Simple mode and Super Simple mode checkboxes are checked Super Simple will be used.

When finished press the Save Modes button.

Simply toggle through the modes on your transmitter and confirm that the PWM for the selected channel matches the required PWM values.

(8)ESC calibration.

Turn on your transmitter and put the throttle stick at maximum.

Connect the Lipo battery. The autopilot's red, blue and yellow LEDs will light up in a cyclical pattern. This means the it's ready to go into ESC calibration mode the next time you plug it in.

With the transmitter throttle stick still high, disconnect and reconnect the battery.

Press and hold the safety button until it displays solid red.

The autopilot is now in ESC calibration mode (On an APM you may notice the red and blue LEDs blinking alternatively on and off like a police car).

Wait for your ESCs to emit the musical tone, the regular number of beeps indicating your battery's cell count (i.e. 3 for 3S, 4 for 4S) and then an additional two beeps to indicate that the maximum throttle has been captured.

Pull the transmitter's throttle stick down to its minimum position.

The ESCs should then emit a long tone indicating that the minimum throttle has been captured and the calibration is complete.

If the long tone indicating successful calibration was heard, the ESCs are "live" now and if you raise the throttle a bit they should spin. Test that the motors spin by raising the throttle a bit and then lowering it again.

Set the throttle to minimum and disconnect the battery to exit ESC-calibration mode.

(9)Set flight mode to "Stabilize".

(10)Arm the motors.

1)Turn on your transmitter.

2)Plug in your LiPo battery. The red and blue lights should flash for a few seconds as the gyros are calibrated (do not move the copter).

3)The pre-arm checks will run automatically and if any problems are found, an APM 2. x will double blink the red arming light, on a Pixhawk the RGB led will blink yellow.

4)Check that your flight mode switch is set to Stabilize, Acro, AltHold, Loiter, or PosHold.

If using a PX4, press the safety button until the light goes solid.

If you are planning on using the autopilot (i. e. Loiter, RTL, Drift, Auto or Guided modes) you should wait for 30 seconds after the GPS has gotten 3d lock. This will give the GPS position time to settle. On APM2 the GPS lock is indicated by the blue LED going solid. On an Pixhawk the RGB LED will blink green.

5)Arm the motors by holding the throttle down, and rudder right for 5 seconds. It takes approximately 5 seconds the first time the copter is armed as it re-initialises the gyros and barometer. Do not hold the rudder right for too long (>15 seconds) or you will begin the AutoTrim feature.

6)Once armed, the red arming light should go solid and the propellers will begin to spin.

(11)Raise the throttle to take-off.

3. 1. 8 Getting Ready to Fly

To prepare to fly, place your copter on your takeoff location, turn your transmitter on and plug in the copter's battery (If you have a PX4 or Pixhawk flight controller depress the start button for five seconds). The ESC should emit a short series of musical notes and then be quiet. You are then ready to arm by holding the throttle stick down and to the right for 5 seconds.

The ESCs and motors automatically disarm after 10 seconds without the motor turning. So anytime you have remained stationary (on the ground) for over 10 seconds you will need to rearm. This is an excellent safety feature. So, after arming you must throttle up the motors within 10 seconds or you will need to re-arm.

When you are done, flying disarm by holding the throttle down and to the left for 5 seconds and disconnect the battery.

A good set of initial PIDs for flying the FlameWheel F330 or F450 in Stabilize Mode (as of Copter version 3. 0. 1) are:

Rate (Roll & Pitch) $P=0.09$ and $I=0.045$ and Stabilize (Roll & Pitch) $P=4.0$.

3.1.9　Conclusions

（1）Choose appropriate standoffs that are suitable for assembly. Don't tighten standoffs with too large force，thus avoiding screw slippery and breakage.

（2）Place the soldering iron on iron stand when soldering iron is not used to solder circuit boards and ESCs temporarily；Power off soldering iron when it is not used for a long time.

（3）Cover the joint between the plug and wire terminals with the Heat-shrink tube.

（4）Watch out for the labels of polarity when soldering leads to the Power Distribution Board.

（5）Remove all propellers before commissioning.

（6）Cover all red points on axis when calibrating compass.

（7）Place quadcopter in correct direction according to the direction on Mission Planner when calibrating accelerometer.

（8）You must set stabilize mode when setting flight mode.

Words & Phrases

sturdy['stɜːdi] *adj.* 坚固的，耐用的

substantial [səb'stænʃl] *adj* 结实的，牢固的

kit[kɪt] *n.* 成套用品；配套元件

bargain['bɑːgən] *n.* 便宜货

rugged['rʌgɪd] *adj.* 结实的

readily['redɪli] *adv.* 轻而易举地；便利地

genuinely ['dʒenjʊɪnli] *adv.* 真诚地；真正地

compatible[kəm'pætəbl] *adj.* 兼容的，相容的

wrap[ræp] *n.* 包裹；包裹物

tie wrap 扎带

assorted[ə'sɔːtɪd] *adj.* 各式各样的

soldering iron 烙铁

rosin cored solder 松香焊锡

thread [θred]（螺纹）locker ['lɒkə(r)] 螺丝胶

Allen wrench [rentʃ] 艾伦(内六角)扳手

lead['liːd] *n.* 引线

sparingly['speərɪŋli] *adv.* 保守地；节约地

excess[ɪk'ses] *n.* 超额量；多余量

keep off （使）不接近

shoulder['ʃəʊldə(r)] *n.* 肩部；肩台

sufficiently[sə'fɪʃntlɪ] *adv.* 足够地,充分地

one at a time 一次一个

tricky['trɪki] *adj.* 微妙的;(形势、工作等)复杂的

off set 排列不正

socket['sɒkɪt] *n.* 插座;灯座

flux[flʌks] *vt.* 熔化;熔解

pad [pæd] *n.* 垫,衬垫

tinned[tɪnd] *adj.* 镀锡的,包锡的

soak up[səuk ʌp] 吸收

out-of-the-box [aʊt] [əv] [ðə] [bɒks] *adj.* 拆盒即可使用的,开箱即用的

optically isolated 光隔离

suspension [sə'spenʃn] *n.* 悬浮;悬架

stick out [stik aʊt] 坚持;伸出来

standoff ['stændɔːf] *n.* 隔离柱,支撑柱

coupled ['kʌpld] *adj.* 联结的,联系的

adhesive [əd'hiːsɪv] *n.* 黏合剂,黏着剂

silicone ['sɪlɪkəʊn] *n.* 硅树脂

electrical[ɪ'lektrɪkl] tape 绝缘带,电线包布

bubble wrap['bʌbl] [ræp] 气泡布

cushioning['kʊʃənɪŋ] *n.* 减震,缓冲

oval ['əʊvl] *adj.* 椭圆形的;卵形的

bore out[bɔː(r)] [aʊt] 镗孔

drill[drɪl] *n.* 钻头

Exercises and Thinking

1. Translate the following sentences.

1) Thread the Tie Wraps from front to back through the frame members and between the motor bullet connector sockets.

2) Use blue or purple removable thread locker very sparingly on the threads of the Allen screws when you assemble the frame.

3) Wrap the battery in bubble wrap and tape it in place to provide additional electrical isolation from the solder joints.

4) It takes approximately 5 seconds the first time the copter is armed as it re-initialises the gyros and barometer.

2. Answer the questions in your own words.

1) What do we need when we assemble a quadcopter?

2) What measures need to be taken to do vibration damping?

3) What's the basic assembly steps of quadcopter?

Mission 2　Fixed-wing Drone Assembly

【Objective】Assemble the Skywalker X8 for myself.

【Analysis】The Skywalker X8 is designed for First-Person-Vision (FPV) application specifically. Most of the material of X8's body is EPO, and the constructional elements are carbon fiber and balsa. It has strong wind resistance to make it fly more stable and high efficiency. The aircraft industrial design is based on computer simulation of automatic wind tunnel modeling. Taking advantage of aerodynamics principle, it has smooth, beautiful body lines as well as stable flying attitude, and it is easy to assembly and disassembly. Equipped with professional placing platform, you can enjoy stable and smooth flying experience.

Skywalker X8 is widely applied in entertainment, sports and scientific research. Stable flying performance gets the favor of practitioner in fields such as land surveying and mapping, aerial photography, rescue and so on.

Skywalker X8 consists of body, winglet, propulsion unit, remote controller, flight controller, mission load and so on (see Fig. 3 - 2 - 1).

The body, located between wings, has a cap on the top and compartment which is mainly used to place airborne electrical device, install mission payload and fix power device.

The wings, on both sides of the body, are the main component that produces lift. And there are winglets that can weaken the induced drag produced by wingtip vortex.

Located at the rear of body, the propulsion unit mainly consists of motor and propeller, drives X8 with the propulsion.

Airborne electrical devices include receiver, flight controller, ESC and so on. They are mainly used to perform the attitude control of X8.

Mission payload, such as electro-optical pod and pan-tilt-zoom camera, can be installed on X8 according to the mission type.

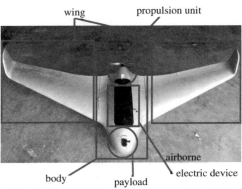

Fig. 3 - 2 - 1　The main components of X8

Table 3 − 2 − 1 shows X8 technical data.

Table 3 − 2 − 1 X8 technical data

Product information		Standard	
Item No.	X8	Motor	Sunnysky4250 KV500/ OS5010 KV810
Wing span	2,120 mm	Propeller	12080F /12060E
Fuselage length	790 mm	Servo	17 g×2 pcs
Wing area	80 dm²	ESC	60A/100A
Flying weight	2,500 − 3,000 g	Battery	22.2V 5,000 − 6,500mAh
C. G.	From head to back 430 − 440 mm	Radio	4CH 4SERVO
Takeoff	Hand cast,catapult shot		
Landing	Glide down,parachute landing		
Maximum anti-wing capability	Level 4		
Air speed	65 − 70 km/h		
Maximum of flying time	25 min		
Maximum of flight level	200 m		

【Knowledge preparation】UAV structure and systems; electromagnetics; circuit; soldering.

3.2.1 What We Need?

KIT version configuration (see Fig. 3 − 2 − 2): fuselage, wing, winglet, sticker, tube and accessories bag which includes glue, screw, layer, rudder horn, extending line, magnets.

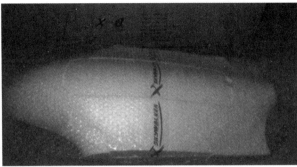

Fig. 3 − 2 − 2 KIT version configuration

Parts listing：carbon fiber tubes，layer，glue，wood chips，screws，rudder horns，steel wire，butt plates，layer board base，magnets，extending lines (see Fig. 3 - 2 - 3).

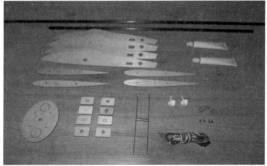

Fig. 3 - 2 - 3　Parts listing

In addition to the equipment introduced above，we also need tools and wear items to complete the assembly，they are listed in Table 3 - 2 - 2.

Table 3 - 2 - 2　Tools listing

Name	Picture	Name	Picture
Digital thermostat welding unit		Lead-free solder wire	
Electrical insulation tape		Fiberglass/ scotch adhesive tape	
Bench vice		Screwdriver KIT	
Wire stripper		Pincer pliers	

Continued

Name	Picture	Name	Picture
Utility knife		Velcro	
Gum brush		Glue gun	
High viscosity glue stick		Servo test	
Long steel tape		Scissors	

3.2.2   Airframe Assembly

1. Glue the winglet layers

As shown in Fig. 3 − 2 − 4 and Fig. 3 − 2 − 5, spread the glue evenly over the wing and both sides of the layers (the layer marked with "T" belongs to parts of the main wing).

Fig. 3 − 2 − 4   Spread the glue evenly over the wing   Fig. 3 − 2 − 5   Spread the glue evenly over the layers

Wait for 5 minutes to allow glue to dry until it is non-sticky, then they can be glued in alignment (see Fig. 3 - 2 - 6).

(a)   (b)

Fig. 3 - 2 - 6   Glue the wing and layer in alignment

2. Mount magnets on the hatch and fuselage

Firstly, compress the magnet tightly, and fix them with CA glue (see Fig. 3 - 2 - 7 and Fig. 3 - 2 - 8).

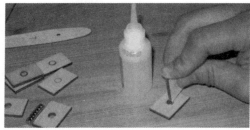

Fig. 3 - 2 - 7   Compress the magnet tightly   Fig. 3 - 2 - 8   Glue the magnet

Secondly, glue the wood chip onto the hatch (see Fig. 3 - 2 - 9) and fuselage (see Fig. 3 - 2 - 10), pay attention to the direction of SN of the magnetic pole.

Fig. 3 - 2 - 9   Glue the wood chip   Fig. 3 - 2 - 10   Glue the wood chip onto

onto the hatch   the hatch and fuselage

After gluing, put it together and inspect the mounting spacing (see Fig. 3 – 2 – 11).

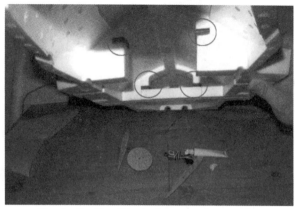

Fig. 3 – 2 – 11   Check the mounting spacing

3. Install the servo

Please choose the standard servo provided by Skywalker or servo with moment more than 2. 5kg, and glue the side face of servo (see Fig. 3 – 2 – 12), then install the servo in place (see Fig. 3 – 2 – 13).

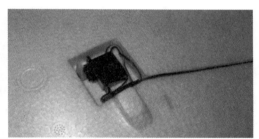

Fig. 3 – 2 – 12   Glue the servo   Fig. 3 – 2 – 13   Installed servo

Spread the glue evenly and wait for its dryness, then glue the rudder horn (see Fig. 3 – 2 – 14) and install the rudder horn in place (see Fig. 3 – 2 – 15).

Fig. 3 – 2 – 14   Glue the rudder horn   Fig. 3 – 2 – 15   Installed rudder horn

Power on the servo and place it in the mid-position, then lock the regulator screw, tighten the steel wire, and shorten the length of the wire (see Fig. 3 – 2 – 16).

Fig. 3 – 2 – 16   Install the steel wire

4. Glue the fixed layer of winglets

Please be sure to spread the glue evenly and wait until it is completely dry, you can glue the layer afterwards (see Fig. 3 – 2 – 17 and Fig. 3 – 2 – 18).

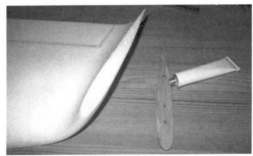

Fig. 3 – 2 – 17   Glue the layer   Fig. 3 – 2 – 18   Glue the winglet

5. Cut out the pre-reserved slot

Cut out the pre-reserved slot on the aileron with utility knife only about 1. 5 mm wide (see Fig. 3 – 2 – 19).

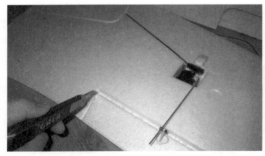

Fig. 3 – 2 – 19   Cut out the pre-reserved   Fig. 3 – 2 – 20   Wing plate bonding
slot on the aileron

6. Wing plate bonding

Please smear evenly on all surfaces, please not to spread the glue over the tube slot (see Fig. 3 – 2 – 20).

Settle the extending line in the line slot previously and spread the glue evenly over the plate (see Fig. 3 - 2 - 21).

Wait for the two planes to be dry, later spread some glue over the vertical planes (see Fig. 3 - 2 - 22), and then fix and flatten them (see Fig. 3 - 2 - 23).

Fig. 3 - 2 - 21   Settle the extending line and spread the glue

Fig. 3 - 2 - 22   Spread some glue over the vertical planes

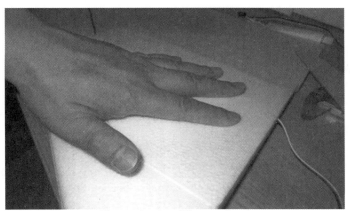

Fig. 3 - 2 - 23   Flatten planes

7. Side plate bonding

Spread the glue on both sides of the plate (see Fig. 3 - 2 - 24).

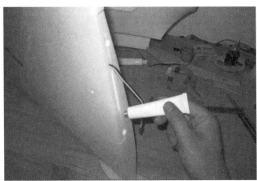 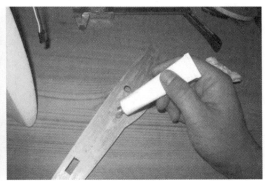

Fig. 3 - 2 - 24   Side plate bonding

It is recommended to insert the tube into the wing to guide the butt plate (see Fig. 3 − 2 − 25), which is more accurate.

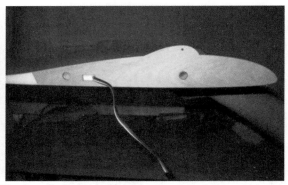

Fig. 3 − 2 − 25　Install the butt plate

8. Widen the line of slot

The line slot of early product also needs to be widened to let the extending line with plug directly into the fuselage (see Fig. 3 − 2 − 26).

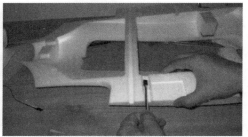

Fig. 3 − 2 − 26　Put the extending line into the slot

9. Fix the motor frame

Cement the motor to the motor frame in advance, you can also first fix the layer board base, and afterwards install motor (see Fig. 3 − 2 − 27).

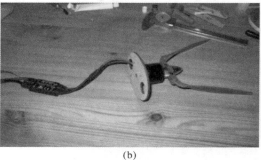

(a)　　　　　　　　　　　　　　　(b)

Fig. 3 − 2 − 27　Fix the motor frame

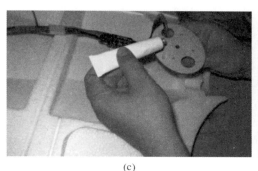
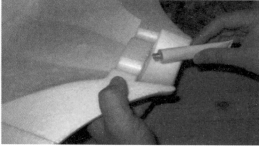

(c)　　　　　　　　　　　　　　　　(d)

Continued Fig. 3 − 2 − 27　Fix the motor frame

10. Align and glue both sides of fuselage

Spread the glue on both sides of the fuselage (see Fig. 3 − 2 − 28), note：please avoid the glue flow into the tube slot！

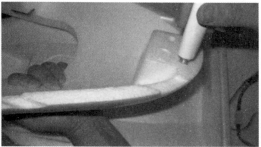

Fig. 3 − 2 − 28　Spread the glue on both sides of the fuselage

After the two sides are dry, align and glue them (see Fig. 3 − 2 − 29).

11. Glue the butt plate

Glue the butt plate of the fuselage the same way as shown in Fig. 3 − 2 − 30.

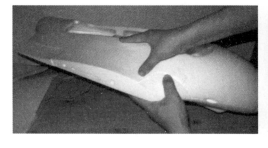

Fig. 3 − 2 − 29　Align and glue two　　　　Fig. 3 − 2 − 30　Glue the butt plate
sides of fuselage

Still use the tube as a guide, align and glue it (see Fig. 3 − 2 − 31).

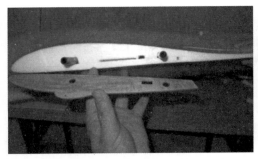

Fig. 3 – 2 – 31　Install the plate

After gluing, you can butt the wing to make a trial of the accuracy, insert the extending into the fuselage through the line (see Fig. 3 – 2 – 32).

We can see the butt is perfect (see Fig. 3 – 2 – 33).

Fig. 3 – 2 – 32　Try to butt the wing　　　　　Fig. 3 – 2 – 33　Perfect butt

As the airframe is designed to be very wide (see Fig. 3 – 2 – 34), you can easily put your FPV equipment down and separate the transmitter from receiver.

12. Fix the wing and winglet

Fixing the wing is also very simple, just a screw (see Fig. 3 – 2 – 35), of course, if you don't trust this construction, you can also use a nut.

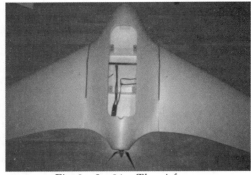
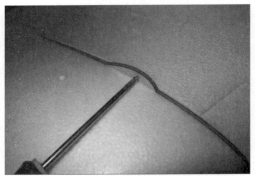

Fig. 3 – 2 – 34　The airframe　　　　　Fig. 3 – 2 – 35　Use a screw to fix the wing

Fix the winglets with two M3×8 screws, with the purpose of a long-term usage of the layer plate. It is suggested that consolidating the screw hole (see Fig. 3 – 2 – 36) with

CA glue before twist the screw (see Fig. 3 - 2 - 37). The well installed winglet is shown in Fig. 3 - 2 - 38.

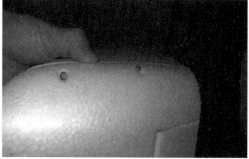

Fig. 3 - 2 - 36   The screw hole

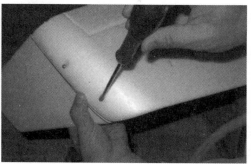

Fig. 3 - 2 - 37   Twist the screw

13. Fix the motor

If you glue the motor frame previously, you can fix the motor now (see Fig. 3 - 2 - 39).

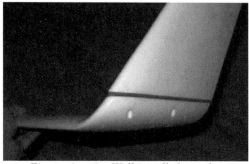

Fig. 3 - 2 - 38   Well installed winglet

Fig. 3 - 2 - 39   Fix the motor

You are suggested to use brand quality propeller of CAM (see Fig. 3 - 2 - 40) or APC.

Fig. 3 - 2 - 40   CAM propeller

14. Mount GoPro type camera or FPV system

You can mount CCD holder and GoPro at the nose of the drone (see Fig. 3 - 2 - 41), the mounting position has already been reserved by the manufacturer.

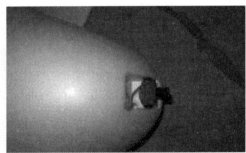

Fig. 3 - 2 - 41   The mounting position of camera

You need to open a round hole in the corresponding position of the GOPRO's lens (see Fig. 3 - 2 - 42). And cut out a bevel to ensure not to obscure the vision of the wide-angle lens.

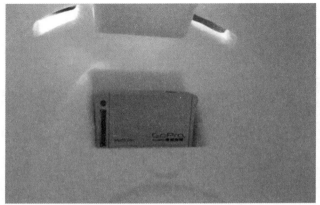

Fig. 3 - 2 - 42   Open a round hole

3.2.3   Airborne Equipment Assembly

1.Connect ESC and motor

Firstly, ensure the receiver is linked with the ID code of the transmitter with which it is being paired; Secondly, connect ESC to motor with three-phase wires, and connect the power, ground and signal wires for ESC to 3rd channel of receiver (see Fig. 3 - 2 - 43); Power on ESC (make sure there is no propeller on drone), push the throttle stick to the highest position, and power on the transmitter afterwards, when the ESC emit "di-di-" beep sound (confirming the maximum throttle), pull the throttle stick to the lowest position in 2s, and the ESC emit "dididi-di" beep sound (confirming the minimum throttle); Thirdly, slightly push throttle stick until the motor is spinning, the rotational direction should be clockwise when looking from tail to nose, if the rotational direction is counterclockwise, just exchange two of the three-phase wires arbitrarily (see Fig. 3 - 2 - 44); Finally, power off the ESC, power off the RC and fix the ESC at the appropriate position of the compartment with Velcro.

Fig. 3 – 2 – 43   Connect ESC and motor

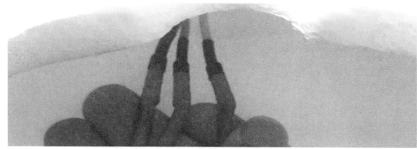

Fig. 3 – 2 – 44   Exchange wires

2. Find the center of gravity and trim the drone

Use the steel tape to measure the position of the center of gravity of X8 (see Fig. 3 – 2 – 45), the result should be 440mm away from the nose with 5mm correction interval in front and at the back of it.

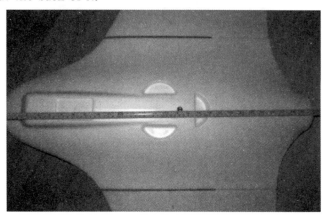

Fig. 3 – 2 – 45   Measure the position of the center of gravity

The drone trim is mainly decided by the installed position of battery. Adjust the installed position of battery with the center of gravity balancing method. When find the appropriate installed position (see Fig. 3 – 2 – 46), fix the battery with Velcro and tie wrap.

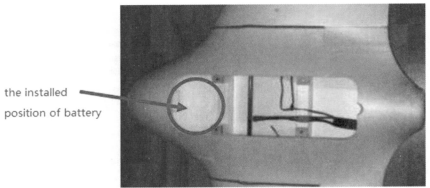

Fig. 3 - 2 - 46   Installed position of battery

3. Install flight controller

We choose the Arkbird Tiny Ⅱ as flight controller for X8. As Skywalker X8 is a kind of all-wing aircraft, we only need ailerons and elevators to control my drone, which means that we need to mix aileron operation with elevator operation.

Connect the output 1 pin of flight controller to right servo, while output 2 pin to left servo and output 3 pin to ESC, and toggle the dial switch to 0 position (see Fig. 3 - 2 - 47).

Fig. 3 - 2 - 47   Connect Arkbird Tiny Ⅱ outputs

Connect output 1, 2, 3, 5 pins of receiver to the input 1, 2, 3, 5 pins of flight controller with female to female 3 wire cables correspondingly (see Fig. 3 - 2 - 47).

Then install receiver, GPS module and flight controller. Firstly, find the appropriate position for them (see Fig. 3 - 2 - 48). Secondly, use Velcro to fix receiver at the planned position and fix GPS with 3M VHB double sides adhesive tape. Finally, install Arkbird Tiny Ⅱ.

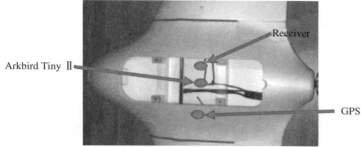

Fig. 3 - 2 - 48   Installed position

There are several matters that should be paid attention to:

(1) The install position of Arkbird Tiny Ⅱ should coincide with center of gravity of X8 (see Fig. 3 - 2 - 49).

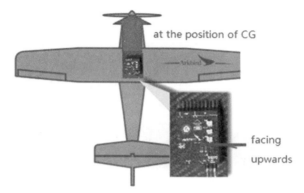

Fig. 3 - 2 - 49   The installed position of flight controller

(2) You can see the printed "FRONT" (see Fig. 3 - 2 - 50) which indicates that the arrow should point the nose direction when installing the flight controller.

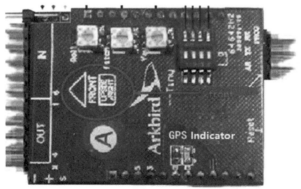

Fig. 3 - 2 - 50   The printed "FRONT"

(3) Place the flight controller at correct attitude according to the instructions.

(4) Fix the flight controller with 3M VHB double sides adhesive tape.

The installed machine can be decomposed into small parts (see Fig. 3 - 2 - 51), which is easy to carry.

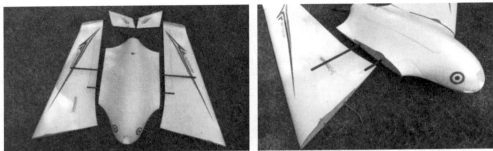

Fig. 3 - 2 - 51　Small parts of X8

Now, we successfully assemble the fixed-wing drone, the full view of our assembly is shown in Fig. 3 - 2 - 52.

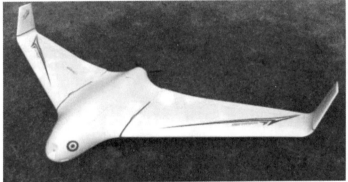

Fig. 3 - 2 - 52　The well assembled X8

3.2.4　Flight Test

After setting the parameters of RC and commissioning the electronic equipment, we can conduct the flight test.

You can try different flight mode and will record the corresponding performance as follows:

(1)Manual mode: the X8 cannot fly at high angle of attack in this mode; The sticks response and the power are normal.

(2)Balancing mode: the X8 can quickly fly with level attitude after releasing the stick.

(3)Fencing mode: normal turning speed, normal hovering altitude, accurate fence position.

(4)Waypoint mode: normal flight path, small GPS positioning error.

According to the flight data and pictures (see Fig. 3 - 2 - 53 and Fig. 3 - 2 - 54), Skywalker X8 works well.

Finally, please note that, control X8's flight speed within 85 km/h. If you need a high-speed flight, please strengthen wings and control the machine load within 3,200g.

Fig. 3 - 2 - 53　Skywalker X8 in flight

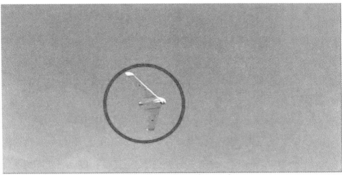

Fig. 3 - 2 - 54　Coordinated turn

3.2.5　Conclusions

The Skywalker X8 is easy to assemble and fly. Just follow the instructions and be careful. There are some tips to watch out for:

(1)Prepare the tools and parts listing well.

(2)Due to the assembly needs many gluing operations, you should practice more to spread the glue evenly on surface and align two surfaces.

(3)As there are many small parts needed to be fixed, you need more patience.

(4)Teamwork is also important in the assembly work, you need to learn how to cooperate with others.

Words & Phrases

winglet ['wɪŋlɪt] *n.* 小翼
glue [ɡluː] *n.* 胶水
moment ['məumənt] *n.* 力矩
hatch [hætʃ] *n.* 舱口
rudder horn['rʌdə(r)] [hɔːn] 舵角
smear [smɪə(r)] *v.* 涂抹
flattern ['flætɜːn] *v.* 压平

butt plate [bʌt pleɪt] 对接板

twist [twɪst] *n.* 拧

consolidate [kənˈsɒlɪdeɪt] *vt.* 巩固

bevel [ˈbevl] *n.* 斜面,斜边

EPO 发泡聚苯乙烯聚乙烯混合体

balsa [ˈbɔːlsə] *n.* 轻木

wingtip vortex [ˈwɪŋ. tɪp] [ˈvɔːteks] 翼尖涡

electro-optical pod [ɪˈlektrəʊ ˈɒptɪkl] [pɒd] 光电吊舱

parts listing 零件清单

consumable material 耗材

wear item 耗材

electrical insulation tape 电绝缘胶布

fiberglass adhesive tape 玻璃纤维胶带

scotch [skɒtʃ] *n.* 格线

bench vice 台虎钳

screwdriver [ˈskruːdraɪvə(r)] *n.* 螺丝刀

pincer pliers 老虎钳

wire stripper 剥线钳

utility knife 美工刀

gum brush 胶水刷

velcro [ˈvelkrəʊ] *n.* 魔术贴

glue gun 胶枪

high viscosity glue stick 高黏胶棒

servo test 舵机调试器

steel tape 卷尺

scissors [ˈsɪzəz] *n.* 剪刀

double sides adhesive tape 双面胶

commission [kəˈmɪʃn] *v.* 调试

Exercises and Thinking

1. Translate the following sentences.

1) Cut out the pre-reserved slot on the aileron with utility knife only about 1.5 mm wide.

2) After gluing, you can butt the wing to make a trial of the accuracy, insert the extending into the fuselage through the line

3) Fix the winglets with two 3×8 screws, with the purpose of a long-term usage of the layer plate, it is suggested that consolidating the screw hole with CA glue before twist the screw.

4)After setting the parameters of RC and commissioning the electronic equipment, we can conduct the flight test.

2. Answer the questions in your own words.

1)What's the procedures of airborne equipment assembly?

2)How to accomplish the airframe assembly?

3)Review the Fixed-Wing Drone Assembly steps, which step is the most difficult? Why?

Mission 3　Helicopter Assembly and Commissioning

【Objective】Build a helicopter for myself.

【Analysis】The assembly of unmanned helicopter is very strict and sophisticated, and it requires certain degree of skill to assemble. Please read the manual of your helicopter carefully before installing and follow all precautions and recommendations located within the manual, the context in this section is only for reference. You are recommended to obtain the assistance of an experienced adult before attempting to assemble the helicopter for the first time. By the act of assembly, you accept all resulting liability.

Ok, now return to the topic, we need to assemble mechanical parts of helicopter precisely (see Fig. 3 – 3 – 1). There are main notes that need to be paid attention to, including mechanical combination between the motor mount and gear, the length of swashplate ball link, the assembling orders of main rotor, the neutral position of mechanical rocker arm of servos, the level of tail holder unit, the adjustment of virtual displacement between fasteners and so on.

Fig. 3 – 3 – 1　Mechanical parts of helicopter

【Knowledge preparation】UAV structure and systems; electromagnetics; circuit; soldering.

3.3.1　Equipment and Tools Required for Assembly

The transmission mode of unmanned helicopter is mainly divided into gear transmission and belt transmission. Before we construct a helicopter, we need to know the transmission mode, then we can check all parts according to the mode. The assembly procedures of helicopters vary from 450 series to 800E. As the helicopter is dangerous and operating it is difficult, we must make an inventory of lists (see Table 3 - 3 - 1) with helicopter, instruction, assembly drawings and product certificate strictly.

Table 3 - 3 - 1　Parts listing

Name	Num	Name	Num
Lateral plate	2	ESC	1
Bottom plate	1	Main blade	1 pair
Battery mount	1	Tail blade	1 pair
Tail boom	1	Tail rotor set	1
Ball linkage rod	4	Servo	4
Swashplate	1	Canopy	1
Motor mount	1	Tail boom brace set	2
Main rotor head	1	Landing gear	1
Slant thread main drive gear set	1	Gyro	1
Motor	1		

The equipment and tools involved in the assembly of helicopter is more sophisticated than other aircraft. So, we should prepare all the equipment and tools listed in Table 3 - 3 - 2 for the assembly.

Table 3 - 3 - 2　Equipment and tools required for assembly

Name	Picture	Name	Picture
Transmitter (6-channel or more, helicopter system)		Philips screw driver $\phi 3.0$, $\phi 1.8$	
Receiver (6-channel or more)		Cutter knife	
Intelligent battery charger		Hexagon screw driver	

Continued

Name	Picture	Name	Picture
22.2 V 6S 4,500 – 5,200 mAh Li-Po Battery ×2 pcs		Needle nose pliers	
Swashplate leveler		CA glue	
Digital pitch gauge		Oil	
Multi-function tester		Grease	

3.3.2 The Whole Machine Assembly

1. MRH assembly

The MRH consists of metal main rotor holders, feathering shaft, spacers, bearings, thrust bearings and screws (see Fig. 3 – 3 – 2).

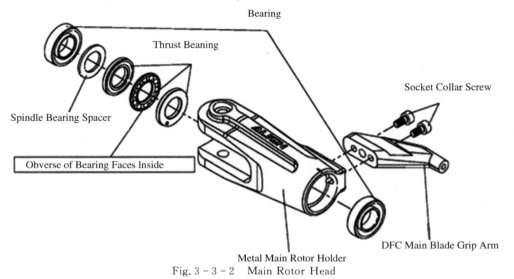

Fig. 3 – 3 – 2   Main Rotor Head

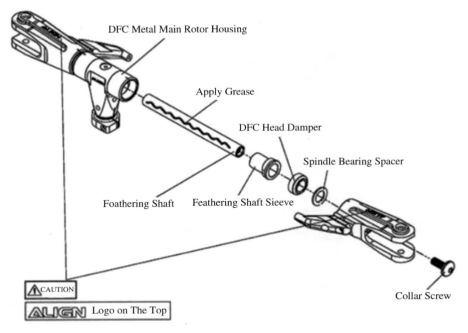

DFC Metal Main Rotor Housing

Apply Grease

DFC Head Damper

Spindle Bearing Spacer

Foathering Shaft

Feathering Shaft Sieeve

⚠ CAUTION

ALIGN Logo on The Top

Collar Screw

Continued Fig. 3 - 3 - 2   Main Rotor Head

Apply grease on thrust bearing and feathering shaft evenly; Put spacer, bearing, metal rotor holder, thrust bearing and screw on feathering shaft at a time (see Fig. 3 - 3 - 3); Then tighten the collar screws on two metal main rotor holders with two screw drivers simultaneously (see Fig. 3 - 3 - 4).

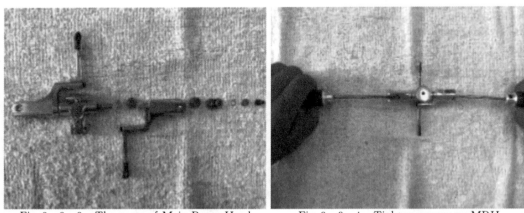

Fig. 3 - 3 - 3   The parts of Main Rotor Head   Fig. 3 - 3 - 4   Tighten screws on MRH

2. Assemble the Main Shaft together with MRH

Distinguish between top end and bottom end of Main Shaft, Insert the top end of Main Shaft into MRH and tighten the screw bolt (see Fig. 3 - 3 - 5).

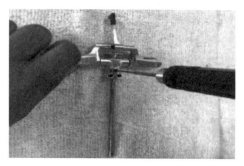

Fig. 3 - 3 - 5   Assemble Main Shaft and MRH

3. Swashplate assembly

Put swashplate on Main Shaft (see Fig. 3 - 3 - 6), then connect the linkage rod precisely measured by vernier caliper (see Fig. 3 - 3 - 7) to the metal main rotor holder. You may adjust the length of ball link when tracking is off in flight.

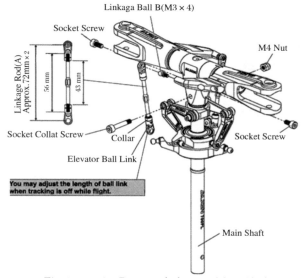

Fig. 3 - 3 - 6   Put swashplate on Main Shaft

Fig. 3 - 3 - 7   Measure the linkage rod

4. Main frames and landing gear assembly

Before assemble the main structure (see Fig. 3 - 3 - 8), please roughly bolt the special collar screw inside the main frame, and then go into next process of main structure assembly. After finishing the main structure assembly, then screw up the special collar screw tightly.

When assembling main frames (see Fig. 3 - 3 - 9), firstly do not fully tighten the screws of main frames and put two bearings through the main shaft to check if the movements are smooth. The bottom bracket must be firmly touched the level table top; please keep the smooth movements on main shaft and level bottom bracket, then slowly tighten the screws. This assembly can influence the power and flight performance.

After assembling the landing gear, install the landing gear on bottom plate (see Fig. 3 - 3 - 10) with screws.

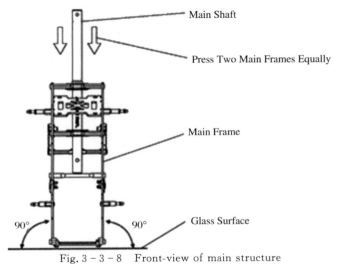

Fig. 3 - 3 - 8   Front-view of main structure

Fig. 3 - 3 - 9   Main frames assembly

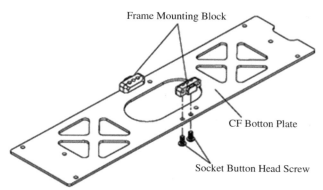

Fig. 3 - 3 - 10   Bottom plate

5. Servos assembly

Install the servos that control the swashplate on main frames without servo arm (see Fig. 3 - 3 - 11), and install the servo that control the tail rotor at the rear of main frames.

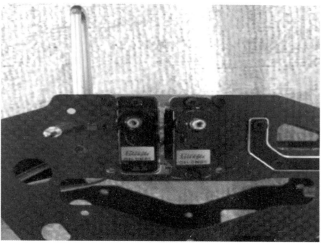

Fig. 3 - 3 - 11   Servos assembly

6. Tail assembly

Fix the tail rotor holder, tail slide sleeve, tail rotor control arm with specialized screws successively according to drawings.

After installing the bearing on metal plate of tail, insert the tail rotor shaft with gear into metal plate, then put the slide shaft and tail rotor holder on tail rotor shaft (see Fig. 3 - 3 - 12) and tighten the screws (see Fig. 3 - 3 - 13).

When assembling umbrella gear, please note to push the gear to the end at a fixed position to make sure the gears mesh with each other smoothly.

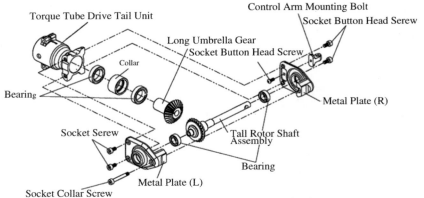

Fig. 3 - 3 - 12   The main structure of tail

After completing the tail rotor assembly, please check if it rotates smoothly.

Fig. 3 - 3 - 13   Tail assembly

7. Motor assembly

Solder banana connectors to motor leads, install gear on motor rotor, fix motor on motor mount without tightening screws up.

When assembling the motor mount, please make sure loose set screw (see Fig. 3 - 3 - 14) on motor gear first, after fully fasten the motor mount with the motor pinion, then fasten back the set screw completely.

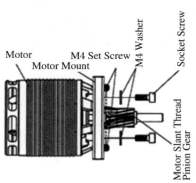

Fig. 3 - 3 - 14   Motor assembly

8. Main rotor and transmission gear assembly

Put the main rotor locking clasp on the main shaft, put transmission gear on the installing position and make it mesh with the pinion gear (see Fig. 3 − 3 − 15), then make main shaft go through the main shaft mount and transmission gear, and lock main shaft on the main shaft mount.

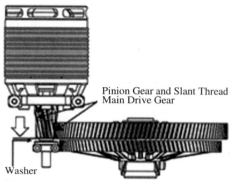

Fig. 3 − 3 − 15   Pinion gear and main gear

The lower edge of main gear needs to be lined up with lower edge of pinion gear (see Fig. 3 − 3 − 16). This will ensure smooth meshing, and avoid interference between pinion's base and main gear which can lead to unusual wear.

Fig. 3 − 3 − 16   Transmission gear assembly

9. Tail tube and tail assembly installation

Insert the tail tube with gearbox into the body, aim at screw-hole and tighten the screws up (see Fig. 3 − 3 − 17).

Fig. 3 − 3 − 17   Tail tube and tail assembly installation

10. Tail boom and stabilizer assembly

When assembling into the tail boom, please apply some oil on the surface, making it smooth during the assembling and keep it vertical with the torque tube for smooth rotation.

In addition, please apply some CA glue to fix bearing on the torque tube, avoid CA glue from the dust causing the bearing stuck. When assembling into the tail boom, please apply some oil and use the attached torque tube mount helper to press the bearing holder of the torque into the tail boom horizontally (see Fig. 3 – 3 – 18).

Fix the horizontal stabilizer and vertical stabilizer at the correct position (see Fig. 3 – 3 – 19), and aim at the screw-hole, tighten the screws up.

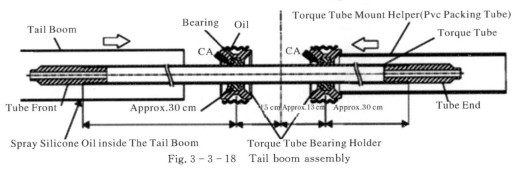

Fig. 3 – 3 – 18　Tail boom assembly

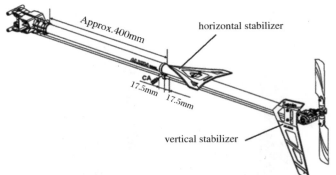

Fig. 3 – 3 – 19　Stabilizer

11. Electronic equipment installation

After soldering banana connector, The ESC can be placed on the main frames (see Fig. 3 – 3 – 20) or bottom plate and need to be fixed with tie wrap; Install the Gyro on the place that connect tail tube with the body of helicopter with the pins towards the tail. Then we can use wires to connect the Gyro with servos and receiver (see Fig. 3 – 3 – 21).

Fig. 3 – 3 – 20　Gyro installation

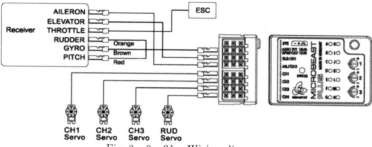

Fig. 3 - 3 - 21　Wiring diagram

12. Canopy assembly

Keep the hole position for canopy mounting bolt horizontally (see Fig. 3 - 3 - 22) to make it easier to insert the R pin to fix the canopy.

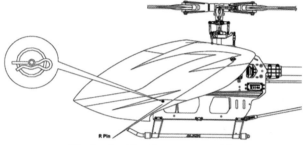

Fig. 3 - 3 - 22　Canopy assembly

13. Considerations

The quality of assembly influences the difficulties of commissioning directly, and bring impact on flight safety. So, we need to take the followings into consideration while assembling our helicopter.

(1)Before installing the servo arm, we need to find the neutral position of servo.

(2)The length of all the linkage rod should be 3 - 5 mm longer than the demand value in order to allow for unforeseen circumstances for helicopter commissioning.

(3)The installing position of Gyro should keep the Gyro away from the magnetic interference as far as possible.

(4)Choose the place with little interference, solidity, high reliability as the installing position of receiver.

(5)Rational wiring can prevent the wiring length shortening and avoid the wires of electric parts to be cut, we need to choose appropriate wiring layout according to the installing position of ESC, Gyro and receiver.

(6)After commissioning, thread lock should be applied to all screws of helicopter, and confirm every screw is firmly secured.

3.3.3　Helicopter Commissioning

1. Mechanical parts commissioning

Mechanical parts commissioning mainly handles the fits between the mechanical parts making the fits of all aspects in relatively perfect condition.

(1)Fine-tune transmission gear.

Fine-tune motor, transmission gear and the length of tail tube, making them work in smooth state. We can fine-tune the motor back and forth by tuning the screws (see Fig. 3 - 3 - 23). You can also make micro adjustments of the length of tail tube according to the tightness of the belt and transmission gear. After these subtle tunings, we can make the gear work smoothly without over tightening that will make gear stuck and over loosing that will cause the slipping of gear.

(2)Find the neutral position of servo.

Connect ESC, servos, gyro and receiver with wires. Un-plug the powerline of motor and power on the helicopter, servo can find the neutral position itself (see Fig. 3 - 3 - 24). Then fix the servo arm and tighten the screws up.

In order to set this option, you need to turn on the transmitter and connect to BEC power.

Note:for the safety, please do not connect ESC to the brushless motor before the setting in order to prevent any accident caused by the motor running during the setting.

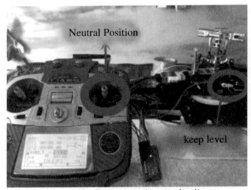

Fig. 3 - 3 - 23　Fine-tune the screws　　　Fig. 3 - 3 - 24　Servo setting and adjustment

(3)Adjustments for gyro and tail neutral setting.

Turn off Revolution mixing mode on the transmitter, then set the gain switch on the transmitter and the gyro to non-head lock mode, or disable gain completely. After setting the transmitter, connect the helicopter power and proceed with rudder neutral point setting. Note:when connecting to the helicopter power, please do not touch tail rudder stick and the helicopter, wait for 3 s to enable gyro , and the rudder servo horn should be 90° to the tail servo (see Fig. 3 - 3 - 25). Tail pitch slider should be half way on the tail output shaft. This will be the standard rudder neutral point. After completing this setting, set the gain switch back to heading lock mode, with gain at around 70%.

After the gyro is enabled and under non-head lock mode, correct setting position of tail servo and tail pitch assembly is as shown Fig. 3 - 3 - 26. If the tail pitch assembly is not in the middle position, please adjust the length of rudder control rod to trim.

Fig. 3 - 3 - 25　Make tail sliding sleeve perpendicular to tail rotor control arm

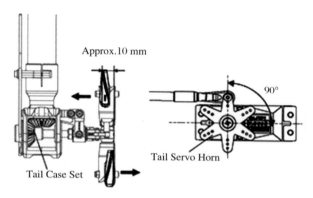

Fig. 3 - 3 - 26　Servo position and tail pitch assembly

2. Remote controller setting

The remote controller check aims to judge whether the helicopter can respond to the signal coming from remote controller. The method of the receiver being linked with transmitter varies from brand to brand. Now we take the futaba remote controller for example.

(1)Follow the servo configuration diagram in Fig. 3 - 3 - 25, plug the servos to Gyro.

(2)Power on the remote controller, unplug the power line that connect ESC to motor.

(3)Press "link mode" key on receiver for 3 s, if the status light of receiver is solid green (see Fig. 3 - 3 - 27), it means that the link between the receiver and remote controller is successfully built.

(4)Power off the helicopter, adjust the throttle stick to the maximum full throttle position; Power on the helicopter, you will hear the highest position acknowledge sound of throttle channel adjustment process; Place the throttle stick to the lowest position, you will hear the lowest position acknowledge sound from ESC. Then we complete the ESC calibration procedure.

(5)Create a new helicopter model.

Select "MODEL TYPE" in the linkage menu and access the setup screen by touching the RTN button.

Select "HELICOPTER" type by scrolling the touch sensor and touch the RTN button. A confirmation message appears. Touch the RTN button for one second. Move to "YES" and touch the RTN button for one second.

(6)Throttle cut setting.

Activate the function: Move the cursor to the "ACT" item and touch the RTN button to switch to the data input mode. Select the ACT mode by scrolling the touch sensor. Touch the RTN button to activate the function and return to the cursor mode.

Switch selection: Move the cursor to the "SW" item and access the switch setup screen by touching the RTN button and select the switch and ON direction.

Throttle cut position setting: Move the cursor to the "POS" item and touch the RTN button to switch to the data input mode. Adjust the servo operation position at throttle cut operation by scrolling the touch sensor. Touch the RTN button to end the adjustment and return to the cursor mode. The setting screen of throttle cut is as shown in Fig. 3 - 3 - 28.

Fig. 3 - 3 - 27　Link is built

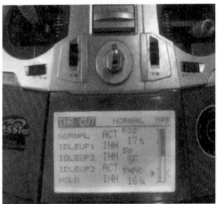

Fig. 3 - 3 - 28　Throttle cut setting

3. Gyro setting

There are different kinds of gyro, and their parameter tuning methods vary. In this section, we will take the tuning method of K-BAR gyro for example.

(1)Turn on the transmitter, power on helicopter, and click the parameter tuning software on computer, connect Gyro to computer with data cable. If the "USB has already been connected" appears at the upper-left position of the software, it demonstrates that the connection is normal.

(2)Determine the main rotor agility and the Gyro gain according to the helicopter type, we can also use the recommended tail yaw ratio and Gyro gain; then we can tune the parameters on the basis of flight condition and feel (see Fig. 3 - 3 - 29).

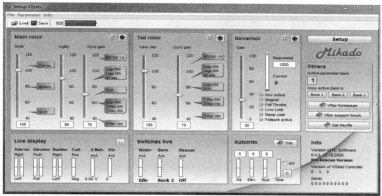

Fig. 3 - 3 - 29　The basic interface

(3)Click the "Install" button on the basic interface, then the menu bar appears; keep the sticks of transmitter at neutral position, click the "set central point" button, then the aileron, elevator, rudder and pitch stay at relative balanced state (see Fig. 3 - 3 - 30).

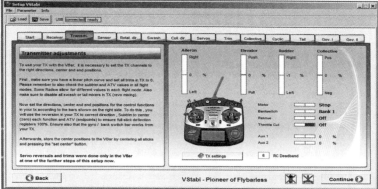

Fig. 3 - 3 - 30　Transmitter setting

(4)We should obey the method displayed in Fig. 3 - 3 - 31 to install the Gyro.

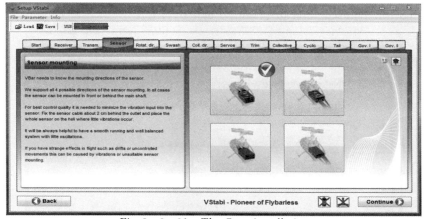

Fig. 3 - 3 - 31　The Gyro installation

(5) The common rotation direction of main rotor is clockwise (see Fig. 3 - 3 - 32), so click the first picture in rotation direction interface.

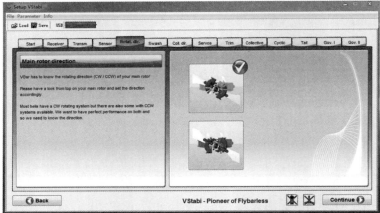

Fig. 3 - 3 - 32　Rotation direction interface

(6) The swashplate tab displays HR-3, H-3, H-4, H-1 types (see Fig. 3 - 3 - 33). As the swashplate type of our helicopter is HR-3, we should select the top-left picture.

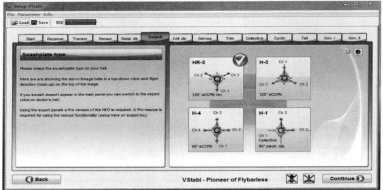

Fig. 3 - 3 - 33　Swashplate types

(7) In pitch direction tab, we choose the upper one (see Fig. 3 - 3 - 34), because it is in accord with the ascending principle of helicopter.

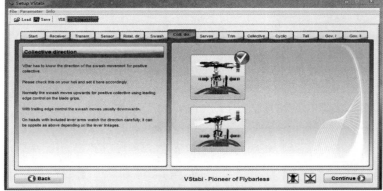

Fig. 3 - 3 - 34　Pitch direction tab

(8)In servo tab, we need to push the throttle stick to judge whether each servo is in the ascending end point (see Fig. 3 – 3 – 35), if not, we need to change the positive and negative end point of servos.

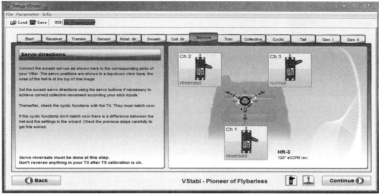

Fig. 3 – 3 – 35   Servo tab

(9)In this tab, we can subtrim the servo. While using flybarless system, please use the swashplate leveler to calibrate swashplate (see Fig. 3 – 3 – 36). Adjust the length of servo linkage rod to make sure the swashplate is leveled before setting up the gyro, thus we can ensure the gyro provides the best performance. Make the main rotor blades be lined up with fuselage, and let the blade wear the pitch gauge in the middle. Adjust the collective pitch until the reading of pitch gauge is 0 (see Fig. 3 – 3 – 37).

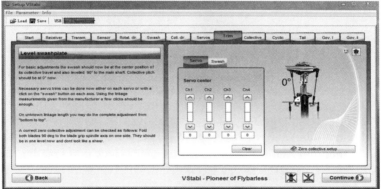

Fig. 3 – 3 – 36   Swashplate calibrating tab

Fig. 3 – 3 – 37   0 pitch

（10）In the collective pitch tab（see Fig. 3 – 3 – 38），push the throttle stick to the highest position，adjust the parameters of transmitter until the value in the pitch tab is between 80 and 100，making the reading of pitch gauge lies between 12 and 14（see Fig. 3 – 3 – 39）.

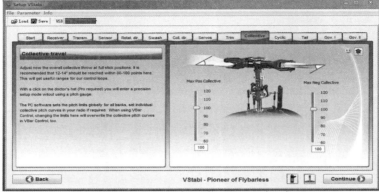

Fig. 3 – 3 – 38　The collective pitch tab

Fig. 3 – 3 – 39　The reading is between 12 and 14

（11）In the cyclic pitch tab（see Fig. 3 – 3 – 40），lay the helicopter flat and put the main rotor blades on the tail tube，the reading of pitch gauge should be 0，click "measure" button, adjust the servo tuning until the reading of pitch gauge changes to 8（see Fig. 3 – 3 – 41）.

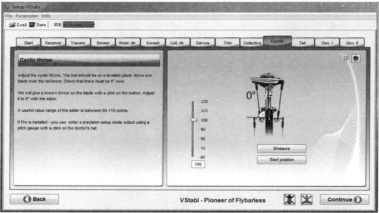

Fig. 3 – 3 – 40　The cyclic pitch tab

Fig. 3 - 3 - 41   The reading is 8

(12)In the tail tab (see Fig. 3 - 3 - 42), toggle the tail servo horn to left (see Fig. 3 - 3 - 43), the tail sliding sleeve will move to right. Adjust the counterclockwise indicator to leave 1 - 2 mm gap between the tail sliding sleeve and the rotor holder. Toggle the tail servo horn to right in the same way, and the adjusting method is similar.

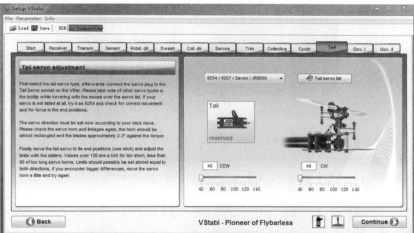

Fig. 3 - 3 - 42   The tail tab

Fig. 3 - 3 - 43   Toggling the tail servo horn

4. Throttle curve and pitch curve setting

(1) Throttle curve setting.

Turn on the transmitter, double click "MDL" key, scroll the cursor to select "THR CURVE" at the Model menu and access the setup screen by touching the RTN button. We can see the screen shown below that the number 5, 4, 3, 2, 1 represent the throttle stick position (see Fig. 3 – 3 – 44) from highest position to lowest position respectively. Beginner is not recommended to set throttle curve to constant. Generally, we set point 1 to 0, point 2 to 40 – 50, and point 3, 4, 5 can be set to the same value which is between 60 and 75. Then scroll the cursor to make the curve between two points smooth, thus when we push the throttle stick over the neutral position, the throttle keep unchanged.

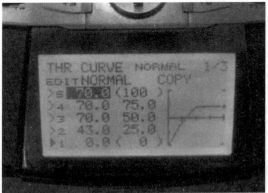

Fig. 3 – 3 – 44   THR CURVE setting screen

(2) Pitch curve setting.

Power on the helicopter, make the blade parallel to tail tube, place the pitch gauge at the middle position of main blade, then double click "MDL" key, scroll the cursor to select "PIT CURVE". We can see the screen shown below that the number 5, 4, 3, 2, 1 represent the pitch stick position from highest position to lowest position respectively. Beginner is recommended to set the pitch value of position 1 and 2 to −2 degrees and 0 degree, and set the pitch value of position 3 and 4 to 7 – 9 degrees, the pitch value of position 5 can not exceed 12 degrees. The curve is shown in Fig. 3 – 3 – 45.

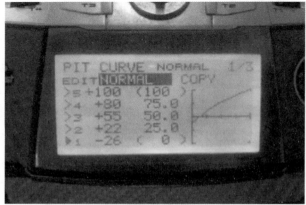

Fig. 3 – 3 – 45   PIT CURVE setting screen

Due to the assembly condition, the servo precision and different motors and ESCs, it's necessary to do flight test and adjust parameters immediately according to the cooperativeness of throttle and pitch.

3.3.4   Conclusions

(1) Read user's manual, instructions, drawings of assembly carefully, Be familiar with brand, type, specifications, characteristics, performance and cautions in detail.

(2) Check all the parts and prepare all tools for assembly, divide the workbench into tools area and parts area tidily, spread out a tablecloth in the middle area and place a box specialized for storing screws.

(3) Generally, the product is divided into several modules, we can check if there exists damage and loss of small parts for each module, after finishing the assembly of each module, we can put them together.

(4) We should assemble main rotor head (MRH) according to the labeled order in drawings, watch out the face direction of the thrust bearing and tighten the bolts of MRH simultaneously with the same torsion in both sides.

(5) Do not tighten screws directly, when there is need to tighten screws simultaneously in both sides, you should obey the principle of tightening screws alternately in "x" shape.

(6) Owing to the strong vibration in flight, you need apply a little amount of threadlocker with 2 – 3 mm width on the threads of screws when you install linkage ball, tail gearbox, main shaft clamping plates, motor mount.

(7) It's necessary to adjust motor mount to ensure that the gear of motor fit well with big gear of helicopter or belt when installing the motor, meshing tightly can cause the extrusion of big gear and the lag of flight which increases electricity consumption; meshing loosely can lead to slipping; press the installed belt to judge whether the mount assembly is good according to the rebounding force of belt.

(8) The length of linkage rod between MRH and swashplate measured by vernier caliper should approach the number in manual to the maximum; and reserve the same amount of screw threads of linkage rod when there are screw threads on both sides.

(9) Use level ruler to mount the tail boom with horizontal installation, it can avoid the force deviation of tail and the fishtailing.

(10) You need to arrange the circuit rationally in order to leave a place where the electromagnetic interference is weak to install the flight controller.

(11) You should connect flight controller correctly after installing the mechanical parts.

(12) Due to the assembly condition, the servo precision and different motors and ESCs, it's necessary to do flight test and adjust parameters immediately according to the cooperativeness of throttle and pitch.

Words & Phrases

swashplate ['swɒʃ'pleɪt] n. 十字盘

swashplate leveler ['lɛvlə] 十字盘校正器

pitch gauge[pɪtʃ] [geɪdʒ] 螺距规

cutter ['kʌtə(r)] knife 美工刀

hexagon screw['heksəgən] [skru:] driver 六角螺丝刀

needle ['ni:dl] nose pliers ['plaɪəz] 尖嘴钳

grease [gri:s] n. 润滑油

tail boom[teɪl] [bu:m] 尾撑

linkage rod['lɪŋkɪdʒ] [rɒd] 连杆

slant thread main drive gear set 斜主齿轮组

tail boom brace [breɪs] set 尾杆支撑组

canopy ['kænəpi] n. 机头罩

thrust bearing[θrʌst] ['beərɪŋ] 止推轴承

gear extrusion[gɪə(r)] [ɪk'stru:ʒn] 齿轮挤压

vernier caliper['vɜ:njə] ['kæləpər] 游标卡尺

screw threads[skru:] [θredz] 螺纹

fishtail ['fɪʃteɪl] v. 摆尾

metal ['metl] main rotor ['rəʊtə(r)] holder ['həʊldə(r)] 金属桨夹

feathering shaft['feðərɪŋ] [ʃɑ:ft] 横轴

spacer ['speɪsə] n. 垫片

slide [slaɪd] shaft 尾轴滑套

banana connector[bə'nɑ:nə] [kə'nektə(r)] 香蕉头

motor mount 电机固定座

motor pinion ['pɪnjən] 电机小齿轮

locking clasp [klɑ:sp] 锁紧扣

horizontal stabilizer[ˌhɒrɪ'zɒntl] ['steɪbəlaɪzə(r)] 水平翼

Exercises and Thinking

1. Translate the following sentences.

1)Before assemble the main structure, please roughly bolt the special collar screw inside the main frame, and then go into next process of main structure assembly.

2)After installing the bearing on metal plate of tail, insert the tail rotor shaft with gear into metal plate, then put the slide shaft and tail rotor holder on tail rotor shaft and tighten the screws.

3)After these subtle tunings, we can make the gear work smoothly without over tightening that will make gear stuck and over loosing that will cause the slipping of gear.

4)In the collective pitch tab, push the throttle stick to the highest position, adjust the parameters of transmitter until the value in the pitch tab is between 80 and 100, making the reading of pitch gauge lies between 12 and 14.

5)The Gel is a material that possess excellent vibration dampening characteristics, ideal for mounting gyroscopes, receivers, or other electronics in vibration prone areas.

6)It uses a folded over 1/16 in O-ring at each corner of the flight control board around a screw sticking out of a small standoff.

7)Taking advantage of aerodynamics principle, it has smooth, beautiful body lines as well as stable flying attitude, and it is easy to assembly and disassembly.

8)There are main notes that need to be paid attention to, including mechanical combination between the motor mount and gear, the length of swashplate ball link, the assembling orders of main rotor, the neutral position of mechanical rocker arm of servos, the level of tail holder unit, the adjustment of virtual displacement between fasteners and so on.

2. Answer the questions in your own words.

1)What kind of equipment and tools are required for assembly?

2)What is the most difficult in the whole machine assembly of helicopter?

3) In the assembly and commissioning of helicopter, what should be paid attention to?

4)How to solder the ESC and battery power leads to the PDB?

5)What equipment and tools do we need to build a electric helicopter?

6)Which procedure do you think is the most difficult in the fixed-wing drone assembly?

【Project Evaluation】

	Work activities	Weightage	Score
	Reading and translation	7	
	Preparation and clearance	3	
	Safe operation	5	
Quadcopter	Frame and motor assembly	10	
	ESC installation and soldering the PDB	10	
	Vibration damping	6	
	Connect ESCs and motors	2	
	Setup ground control station	2	
	RC settings	2	
	Sensors calibration	8	

Continued

	Work activities	Weightage	Score
Fixed-wing Drone	Fuselage assembly	2	
	Assemble control surfaces	5	
	Install servos	4	
	Install motor and fix propeller	4	
	Connect ESC and motor	1	
	Calibrate the centre of gravity	2	
	Install the flight controller	2	
Helicopter	MRH assembly	1	
	Swashplate assembly	4	
	Servos and motor assembly	2	.
	Tail assembly	2	
	Main rotor and transmission gear assembly	2	
	Electronic equipment installation	3	
	Mechanical parts commissioning	2	
	RC and Gyro settings	4	
Attitude and attendance		5	

Grade:□Excellence □Good marks □Medium level □Pass

【Project Conclusions】 Upon completion of this project，you shall be able to：

1)Assemble and commission a quadcopter alone；

2)Assemble an electric fixed-wing drone；

3)Follow the manual to construct an electric helicopter.

Project 4　Fly Your Drone

【Description】 This project will help you to master the flight dynamics of UAV, and you can learn how to fly your drone.

【Analysis】 You are recommended to preview the key words and missions in advance and take notes, then listen carefully in class and participate in discussion actively, and furthermore, you should follow the instructions to do a lot of flight practice. Diligence plus correct striving direction can bring you the best reward.

【Related knowledge and skills】 Aeronautical Conspectus; UAV Regulations.

Read and translate some long and difficult sentences; know how to use remote controller.

Mission 1　The Flight Dynamics of UAV

【Objective】 Master the flight principle of UAV; know how the aerodynamic forces act on the fixed-wing drone and how the control surfaces work upon the fixed-wing drone; know how the helicopter stay in the air and steer; understand the flight dynamics of quadcopter.

【Analysis】 Flight dynamics of UAV refers to the study of the performance, stability, and control of unmanned aerial vehicles. It is concerned with how forces acting on the vehicle influence its speed and attitude with respect to time. Mastering the flight dynamics help us to control the fight of UAV and fly it well.

【Knowledge preparation】 Three laws of mechanics.

Flight dynamics of UAV refers to the study of the performance, stability, and control of unmanned aerial vehicles. It is concerned with how forces acting on the vehicle influence its speed and attitude with respect to time.

4.1.1　Fixed-wing Drone

The three critical flight dynamics parameters of fixed-wing drone are the angles of rotation in three dimensions about the vehicle's center of mass, known as pitch, roll and yaw (see Fig. 4 − 1 − 1).

The most common aeronautical convention defines the roll as acting about the longitudinal axis, positive with the starboard (right) wing down [see Fig. 4 - 1 - 2(a)]. The yaw is about the vertical body axis, positive with the nose to starboard [see Fig. 4 - 1 - 2(c)]. Pitch is about an axis perpendicular to the longitudinal plane of symmetry, positive nose up [see Fig. 4 - 1 - 2(b)].

A fixed-wing drone increases or decreases the lift generated by the wings when it pitches nose up or down by increasing or decreasing the Angle of Attack(AOA). The roll angle is also known as bank angle on a fixed-wing aircraft, which usually "banks" to change the horizontal direction of flight. An aircraft is usually streamlined from nose to tail to reduce drag making it typically advantageous to keep the sideslip angle near zero.

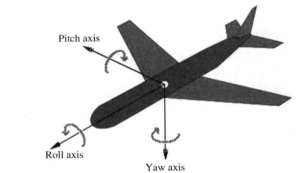

Fig. 4 - 1 - 1   The rotation in three dimensions

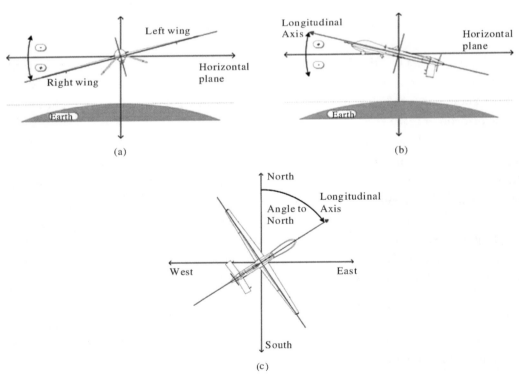

Fig. 4 - 1 - 2   Angles definition

(a)roll angle definition; (b)pitch angle definition; (c)yaw or heading angle definition

1. Aerodynamic forces

Essentially there are 4 aerodynamic forces that act on an airplane in flight; these are lift, drag, thrust and weight (see Fig. 4 – 1 – 3).

In simple terms, drag is the resistance of air molecules hitting the airplane (the backward force), thrust is the power of the airplane's engine (the forward force), lift is the upward force and weight is the downward force. So, for airplanes to fly and stay airborne, the thrust must be greater than the drag and the lift must be greater than the weight.

This is certainly the case when an airplane takes off or climbs. However, when it is in straight and level flight, the opposing forces of lift and weight are balanced. During a descent, weight exceeds lift; and to slow, an airplane drag has to overcome thrust.

The picture below shows how these 4 forces act on an airplane in flight:

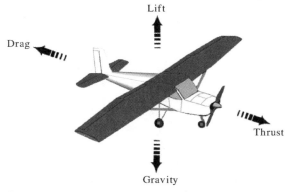

Fig. 4 – 1 – 3　Aerodynamic forces on fixed-wing drone

Thrust is generated by the airplane's engine (propeller or jet), weight is created by the natural force of gravity acting upon the airplane and drag comes from friction as the plane moves through air molecules. Drag is also a reaction to lift, and this lift must be generated by the airplane in flight. This is done by the wings of the airplane.

The generation of lift has been an argued theory in the past, but certain principles have been known about and agreed on for a long time now.

A cross section of a typical airplane wing show the top surface to be more curved than the bottom surface (see Fig. 4 – 1 – 4). This shaped profile is called an "airfoil" and the shape exists because it's long been proven (since the dawn of flight) that an airfoil generates significantly more lift than opposing drag.

During flight, air naturally flows over and beneath the wing and is deflected upwards over the top surface and downwards beneath the lower surface. Any difference in deflection causes a difference in air pressure ("pressure gradient") and because of the airfoil shape the pressure of the deflected air is lower above the airfoil than below it. As a result, the wing is "pushed" upwards by the higher pressure beneath.

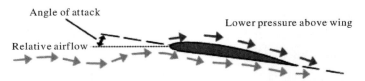

Fig. 4 - 1 - 4　The general movement of air over an airfoil

The faster a wing moves through the air, so the actions are exaggerated and more lift is generated. Conversely, a slower moving wing generally creates less lift.

It's important to note, though, that different wing designs (airfoil and shape) generate lift more (and less) efficiently than other designs at different speeds, depending on what the plane has been designed for.

2. Control surfaces

For an airplane to fly in a controlled manner, control surfaces are necessary. The 4 main surfaces (see Fig. 4 - 1 - 5) are ailerons, elevators, rudder and flaps as shown below:

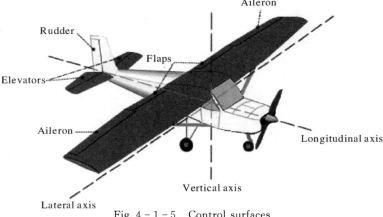

Fig. 4 - 1 - 5　Control surfaces

To understand how each works upon the airplane, imagine 3 lines (axis— the blue-dashed lines in the picture above) running through the plane. One runs through the centre of the fuselage from nose to tail (longitudinal axis), one runs from side to side (lateral axis) and the other runs vertically (vertical axis). All 3 axes pass through the Centre of Gravity (CG), the airplane's crucial point of balance.

When the airplane is in forward flight, it will rotate around each axis when movement to any control surface is made by the operator or flight controller.

The following sections explain how each control surface effects the airplane. It's important to understand that all control surfaces work in the same way, in that they alter the camber (airfoil shape)of the complete flying surface. This, in turn, changes the forces acting on the surfaces and so that surface reacts in accordance with the change in force. The force in question is best known to us as lift, but this particular force occurs in any direction-not just upwards.

3. Ailerons

Located on the trailing edge (rear) of the wing, the ailerons control the airplane's roll about its longitudinal axis. Each aileron moves at the same time but in opposite directions i. e. when the left aileron moves up, the right aileron moves down and vice versa (see Fig. 4 - 1 - 6).

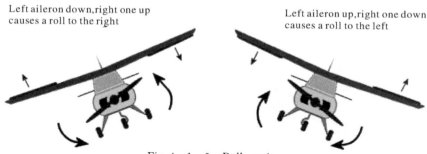

Left aileron down,right one up causes a roll to the right

Left aileron up,right one down causes a roll to the left

Fig. 4 - 1 - 6　Roll motion

This movement causes a slight decrease in lift on the wingtip with the upward moving aileron, while the opposite wingtip experiences a slight increase in lift. Because of these subtle changes in lift, the airplane is forced to roll in the appropriate direction i. e. when the operator moves the control stick to the left on the transmitter, the left aileron will rise and the airplane will roll left in response to the change in lift on each wing.

4. Elevators

The elevators are located on the rear half of the tailplane, or horizontal stabiliser. The job of the tailplane is to generate a downward force to counteract the natural nose-diving tendency of planes, which happens as a result of the natural forces that are generated about a plane's Centre of Gravity and Centre of Lift.

As the elevators are deflected up or down, so the amount of down force changes and this results in the airplane's nose pitching up or down; up elevator means more down force, so the plane pitches up, and vice versa (see Fig. 4 - 1 - 7).

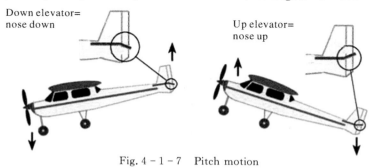

Down elevator= nose down

Up elevator= nose up

Fig. 4 - 1 - 7　Pitch motion

However, pitching the nose up doesn't necessarily mean the plane will climb. In fact, it's quite possible to be flying level, or even descending, with a nose-up attitude. Only when power is added and speed increased, will the plane climb with up elevator.

Elevators are the single most important control surface of a plane, and they effect the airplane's airspeed more than the need to climb or dive.

5. Rudder

The rudder makes up the rear portion of the vertical stabiliser, or fin. When the operator moves the rudder stick to the left on the transmitter, the rudder moves to the left, while depressing the rudder stick to the right deflects the rudder to the right (see Fig. 4 - 1 - 8).

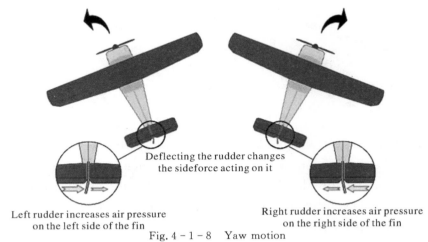

Deflecting the rudder changes the sideforce acting on it

Left rudder increases air pressure on the left side of the fin

Right rudder increases air pressure on the right side of the fin

Fig. 4 - 1 - 8　Yaw motion

The rudder works in the same way as ailerons and elevators, in that it changes the airflow over the fin.

Essentially, you can think of a fin as a vertical wing. The air flowing over it and the rudder acts in exactly the same way as it does flowing over a wing and aileron-except the forces are vertical and not horizontal.

Deflecting the rudder to the left increases the air pressure on the left side of the fin and rudder, and so the whole back end of the plane is pushed across to the right, thus yawing the nose to the left.

6. Flaps

Flaps are located on the trailing edge of each wing, usually between the fuselage and the ailerons, and extend downward (and often outward) from the wing when put into use. The purpose of the flaps is to generate more lift at slower airspeed, which enables the airplane to fly at a greatly reduced speed with a lower risk of stalling. When extended further flaps also generate more drag which slows the airplane down much faster than just reducing throttle (see Fig. 4 - 1 - 9).

Although the risk of stalling is always present, generally speaking an airplane has to be flying very slowly to stall when flaps are in use at, for example, 10 degrees deflection. Obviously though stall speeds and safe airspeeds vary from airplane to airplane.

So, all these factors are why and how airplanes fly. Radio Control (RC) planes can of course be more simple—for example, just have rudder and elevator control or perhaps just rudder and motor control.

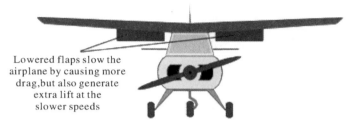

Fig. 4 – 1 – 9　Yaw motion

4. 1. 2　Helicopter

Helicopters truly are amazing aircraft. How helicopters fly is what makes them such versatile machines, being perfectly suited to roles ranging from military use to firefighting to Search and Rescue (SAR), and everything in between.

Yamaha developed its first industrial-use unmanned helicopter, the R-50, in 1987, after a request from the Ministry of Agriculture, Forestry and Fishery of Japan. Unmanned helicopter for agriculture has the ability to increase productivity, whilst reducing the risks surrounding the use of traditional farming equipment in adverse conditions. Some unmanned helicopters are designed for a wide range of industrial and research applications. Some can be used for photography, surveillance, photogrammetry and information gathering.

A typical unmanned helicopter has thousands of intricate components, but we only need to worry about a handful of the bigger bits (see Fig. 4 – 1 – 10). The main framework is called the fuselage. It contains one or two engines, a transmission, and gearboxes, which power one or two main rotors and a much smaller tail rotor at the back.

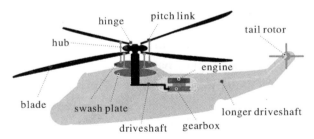

Fig. 4 – 1 – 10　The essential, mechanical parts of a helicopter

Each rotor blade is connected to the hub and rotating mast by a feathering hinge, which allows it to swivel. A pitch link (a short rod) attached to each blade can tilt it to a steeper or shallower angle according to the position of the rotating upper swash plate, which spins on bearings around the static lower swash plate. That's how a chopper hovers and steers. The two swash plates are moved up and down or tilted to the side by the operator's cyclic and collective controls. The rotor is powered by a driveshaft connected to a transmission and gearbox. The same transmission powers a second, longer driveshaft connected to a gearbox that spins the tail rotor. The power from both rotors comes from one or two engines.

A normal airplane can fly forwards, climb upwards, dive downwards and turn (and roll)to the left and right. A helicopter can do all this plus has the ability to fly backwards, move sideways in any direction, rotate 360 degrees on the spot and hover (see Fig. 4 – 1 – 11)i. e. stay airborne with no directional movement at all.

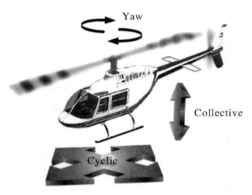

Fig. 4 – 1 – 11　The directions and the associated name of control

1. How does a helicopter stay in the air?

The science of a helicopter is exactly the same as the science of an airplane: it works by generating lift—an upward-pushing force that overcomes its weight and sweeps it into the air. Planes make lift with airfoils (wings that have a curved cross-section). As they shoot forwards, their wings change the pressure and direction of the oncoming air, forcing it down behind them and powering them up into the sky: a plane's engines speed it forward, while its wings fling it up. The big problem with a plane is that lots of air has to race across its wings to generate enough lift; that means it needs large wings, it has to fly fast, and it needs a long runway for takeoff and landing.

Helicopters also make air move over airfoils to generate lift, but instead of having their airfoils in a single fixed wing, they have them built into their rotor blades, which spin around at high speed (roughly 500 r/min, revolutions per minute). The rotors are like thin wings, "running" on the spot, generating a massive downdraft of air that blows the helicopter upward. With skillful piloting, a helicopter can take off or land vertically, hover or spin on the spot, or drift gently in any direction—and you can't do any of that in a conventional plane.

2. How does a helicopter hover and steer?

A helicopter's rotors are ingenious things that allow it to hover in mid – air or steer in any direction.

(1)Hovering.

As they start to spin around, the airfoils on the rotor blades generate lift that over-comes the weight of the craft, pushing it up into the air. If the lift is greater than the weight, the helicopter climbs; if it's less than the weight, the helicopter falls. When the lift and the weight are exactly equal, the helicopter hovers in mid-air. The operator can make the rotor blades generate more or less lift using a control called the collective pitch (see Fig. 4 – 1 – 12), which increases or decreases the angle ("pitch") that all the blades make to the oncoming air as they spin around. For takeoff, the blades need to make a steep angle to generate maximum lift.

Fig. 4 – 1 – 12　Collective pitch

The collective pitch control changes the angle (or pitch) of each of the rotor blades by the same amount at the same time (green arrows)—in other words, collectively. If the blades make a steeper angle, they generate more lift so the entire craft moves straight upward (orange arrow).

How does that happen? As we've already seen, the main rotor is connected to the hub at the top of the mast by a feathering hinge that allows each blade to swivel as it spins, so it makes a steeper or shallower angle to the oncoming air. The blades have short vertical rods (pitch links) attached to them that are connected to a rotating metal disc called a swash plate, a bit lower down the mast. This swash plate slides on bearings around a second, similar plate directly underneath that doesn't rotate. When the opera-tor moves the collective one way, both swash plates move upward, pushing up on the pitch links that tilt the rotor blades to a steeper angle. Moving the collective the other way moves the swash plates back down, pulling on the pitch links and tilting the blades to a shallower angle.

　(2) Steering.

The rotors also provide the steering for a helicopter by making more lift on one side than the other. They do this by swiveling back and forth (feathering) as they rotate, so, for example, they make a steeper angle when they're on the left side of the craft than when they're on the right. That means they generate more lift on the left, tilting the craft over to the right and steering it in that direction. The operator steers like this using a second lever called the cyclic pitch (also known as the "cyclic stick" or just "cyclic"), similar to a joystick, which makes the blades swivel as they cycle around. The cyclic pitch control changes the angle of selective rotor blades as they spin, so (in this case) whichever blade is on the left always produces slightly more lift, while the opposite blade (shown here on the right)always produces slightly less lift (see Fig. 4 – 1 – 13). That means more lift is produced on the left side of the helicopter, so the overall lift (orange arrow)is tilted to the right, steering the entire helicopter in that direction.

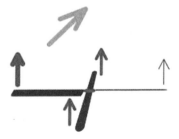

Fig. 4 – 1 – 13　Cyclic pitch

The ingenious swash plate mechanism translates the operator's commands into appropriate movements of the rotor blades. As shown in Fig. 4 – 1 – 14, you can see a simplified view of the swash plate mechanism. There are two discs at the top of the rotor mast, an upper one (red)that rotates on ball bearings (orange)around a lower one (blue)that doesn't rotate at all. Four pitch links (green)connect the upper swash plate to the rotor blades. Suppose the operator wants to fly to the right. First, the command moves the cyclic to the right. That tilts both swash plates over to the right. As the rotor blades rotate, the tilted swash plates force the pitch links up when they're on the left and down when they're on the right. That makes each rotor blade tilt to a steeper angle when it's on the left and a shallower angle when it's on the right. This produces more lift on the left, steering the chopper to the right.

Fig. 4 – 1 – 14　Swash plate mechanism

The operator can also steer the nose of a helicopter in a certain direction using yaw control stick, which changes the pitch of the tail rotor blades so they make more or less sideways thrust than in normal straight flight. That makes the entire craft rotate slowly clockwise or counterclockwise so it heads in a different direction.

4.1.3　Quadcopter

Each rotor produces both a thrust and torque about its center of rotation, as well as a drag force opposite to the vehicle's direction of flight. If all rotors are spinning at the same angular velocity, with rotors one and three rotating clockwise and rotors two and four counterclockwise, the net aerodynamic torque, and hence the angular acceleration about the yaw axis, is exactly zero, which means there is no need for a tail rotor as on conventional helicopters. Yaw is induced by mismatching the balance in aerodynamic torques (i. e. , by offsetting the cumulative thrust commands between the counter-rotating blade pairs).

Fig. 4 - 1 - 15 shows the direction of travel for each propeller respective to the direction of travel. A and C propellers are moving clockwise while B and D propellers are moving counterclockwise.

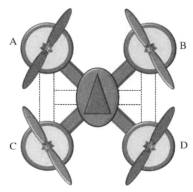

Fig. 4 - 1 - 15   Direction of travel for each propeller

Spinning rotors can produce reaction torques on each motor of a quadcopter aircraft. Rotors A and C spin in one direction, while rotors B and D spin in the opposite direction, yielding opposing torques for control.

Fig. 4 - 1 - 16(a) shows that a quadrotor hovers or adjusts its altitude by applying equal thrust to all four rotors.

Fig. 4 - 1 - 16(b) shows that a quadrotor adjusts its yaw by applying more thrust to rotors rotating in one direction.

Fig. 4 - 1 - 16(c) shows that a quadrotor adjusts its pitch or roll by applying more thrust to one rotor and less thrust to its diametrically opposite rotor.

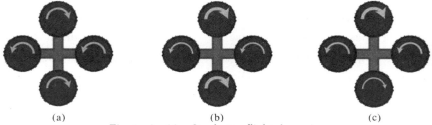

(a)   (b)   (c)
Fig. 4 - 1 - 16   Quadrotor flight dynamics

1. Coaxial configuration

In order to allow more power and stability at reduced weight, a quadcopter, like any other multirotor, can employ a coaxial rotor configuration (see Fig. 4 - 1 - 17). In this case, each arm has two motors running in opposite directions (one facing up and one facing down).

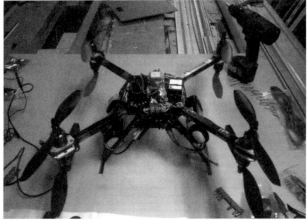

Fig. 4 - 1 - 17　Coaxial Quadcopter

2. Mechanical structure

The main mechanical components needed for construction are the frame, propellers (either fixed-pitch or variable-pitch), and the electric motors. For best performance and simplest control algorithms, the motors and propellers should be placed equidistant. Recently, carbon fiber composites have become popular due to their light weight and structural stiffness.

The electrical components needed to construct a working quadcopter are similar to those needed for a modern RC helicopter. They are the electronic speed control module, on-board computer or controller board, and battery. Typically, a hobby transmitter is also used to allow for human input.

3. Autonomous flight

Quadcopters and other multicopters often can fly autonomously. Many modern flight controllers use software that allows the user to mark "way-points" on a map, to which the quadcopter will fly and perform tasks, such as landing or gaining altitude. The PX4 autopilot system, an open-source software/hardware combination in development since 2009, has since been adopted by both hobbyists and drone manufacturing companies alike to give their quadcopter projects flight-control capabilities. Other flight applications include crowd control between several quadcopters where visual data from the device is used to predict where the crowd will move next and in turn direct the quadcopter to the next corresponding waypoint.

4.1.4　Conclusions

(1)The 4 main surfaces to fly a fixed-wing drone in a controlled manner are ailerons, elevators, rudder and flaps.

（2）We can alter the camber of the complete flying surface to change the attitude of fixed-wing drone by moving sticks of RC.

（3）You need use the collective pitch control to make the helicopter hover in the sky, and the cyclic pitch control changes the angle of selective rotor blades to steer the helicopter.

（4）You can change the attitude of quadcopter by changing the speed of rotors. Sometimes, you need to change two of them; and you need to change the speed of all rotors at other times.

Words & Phrases

swivel['swɪvl] *vt.*（使）旋转；在枢轴上转动

perpendicular [ˌpɜːpən'dɪkjələ(r)] *adj.* 垂直的，成直角的

sideslip ['saɪdslɪp] *n.* 侧滑

profile['prəʊfaɪl] *n.* 侧面；外形，轮廓

airfoil ['eəfɔɪl] *n.* 翼面

deflect[dɪ'flekt] *vt.* 使弯曲；偏转

camber['kæmbə(r)] *n.* 拱形

versatile ['vɜːsətaɪl] *adj.* 多功能的；多用途的

hinge [hɪndʒ] *n.* 铰链

gearbox['ɡɪəbɒks] *n.* 变速箱；齿轮箱；变速器

collective pitch [kə'lektiv] [pitʃ] 总距

rod [rɒd] *n.* 杆，拉杆

cyclic pitch ['saɪklɪk] [pitʃ] 周期桨距

bearing['beərɪŋ] *n.* 轴承

Exercises and Thinking

1. Translate the following sentences.

1）Any difference in deflection causes a difference in air pressure （"pressure gradient"）and because of the airfoil shape the pressure of the deflected air is lower above the airfoil than below it.

2）Helicopters also make air move over airfoils to generate lift, but instead of having their airfoils in a single fixed wing, they have them built into their rotor blades, which spin around at high speed （roughly 500r/min, revolutions per minute）.

3）The cyclic pitch control changes the angle of selective rotor blades as they spin, so （in this case）whichever blade is on the left always produces slightly more lift, while the opposite blade （shown here on the right）always produces slightly less lift.

4）If all rotors are spinning at the same angular velocity, with rotors one and three rotating clockwise and rotors two and four counterclockwise, the net aerodynamic torque, and hence the angular acceleration about the yaw axis, is exactly zero, which means there is no need for a tail rotor as on conventional helicopters.

2. Answer the questions in your own words.

1) Why the wing of an airplane can produce lift?

2) How does the helicopter hover in mid-air?

3) How does the helicopter steer?

4) Describe the flight dynamics of quadcopter.

Mission 2　Flying with Simulator

【Objective】Learn about what it takes, in terms of focus, clear thinking, and planning, to manage a flight and land without crashing. You will get a good sense of what it really takes to learn to control a drone.

【Analysis】Flight simulators are used extensively for UAV pilot training, especially for complex and very expensive aircraft. While obviously they cannot substitute for real flying, they do provide excellent learning opportunities for a very low cost and total safety.

You are recommended to use Phoenix R/C, which is the professional model flight simulator for all levels of R/C pilot-from absolute beginner to seasoned competition-level flier. You can get sophisticated control mechanisms like autorotation, hovering, takeoff, landing and so on.

You will get tremendous help by carefully, and methodically, using a flight simulator. Set the realism to as high as your computer can manage, and perform your flights with the same care and discipline you would use in a real drone flight. Go from engine startup, taxi, cruise and navigation, descent, approach, and landing to full stop with fixed-wing drone, and from arming motors, pushing throttle, hovering, autorotation, tail-in hovering, nose-in hovering to full stop with rotary drone.

You'll be able to take the controls under supervision and feel the exhilaration of flight.

【Knowledge preparation】Computer applications; RC operation.

4.2.1　The Correct Hand-held Method of Remote Controller

1. Move the stick with single fingertip

This method uses two thumb fingertips to move the left and right sticks, the index fingers and middle fingers are used to toggle switches, while the little fingers and ring fingers are responsible for holding the body of remote controller (see Fig. 4-2-1). This method is easy for us to master with the advantage of strong agility of controlling sticks.

Fig. 4 - 2 - 1   Move the stick with single fingertip

2. Move the stick with two fingertips

This method uses two thumb fingertips to move the sticks mainly, and the index fingers are placed upon the sticks functioning as auxiliary control, the middle fingers and ring fingers are used to toggle switches, while the little fingers are responsible for holding the body of remote controller (see Fig. 4 - 2 - 2). This method has high control accuracy of sticks with good stability and can lower the influence of low accuracy arose by tiredness due to long time repeated finger movement. So, this method is suitable for beginners.

Fig. 4 - 2 - 2   Move the stick with two fingertips

4. 2. 2   Japanese Hand and American Hand

1. American hand

The left stick of the remote controller controls how much lift your drone is creating which allows it to ascend and descend, and rotates the drone around its center either clockwise or counterclockwise (see Fig. 4 - 2 - 3). The right stick controls the forward and backward movement of your drone, and moves the drone horizontally left and right (see Fig. 4 - 2 - 3).

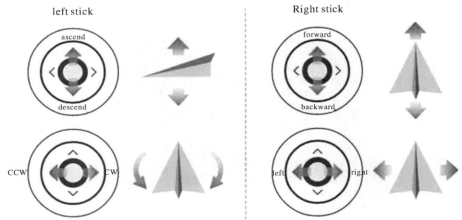

Fig. 4 - 2 - 3 American hand

2. Japanese hand

The difference between the Japanese hand and the American hand is actually not that big, only that the control of the drone's ascend / descend and forward / backward were reversed. That is, the left joystick is responsible for the forward / backward movement of the drone, and clockwise / counterclockwise rotation on the spot (see Fig. 4 - 2 - 4); while the right joystick controls the rise / fall of the drone and left / right movement (see Fig. 4 - 2 - 4).

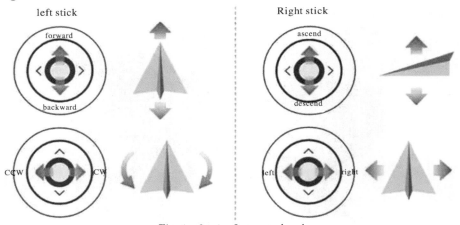

Fig. 4 - 2 - 4 Japanese hand

4.2.3 Auxiliary Exercise with Remote Controller

Fingering exercises are applicable to the preliminary stage. In order to improve the precision of flight operation, we need to proceed auxiliary exercise with remote controller that is exercising the precision of each channel under the supervision of "SERVO MONITOR" screen (see Fig. 4 - 2 - 5). The demand is move one stick slowly with the position value from −100 to 100, and the other channels remain unchanged at the same time.

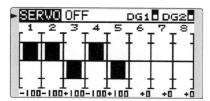

Fig. 4 - 2 - 5 SERVO MONITOR screen

4.2.4 Using Phoenix R/C

Phoenix features over 200 models out of the box, each built and tuned by established industry veterans and in most cases by the actual model designers and developers themselves for the most accurate simulation possible. Phoenix also features a great range of training and competition modes which makes learning to fly easy and fun. These modes will take you through from your first Hover to advanced manoeuvres such as autorotations.

1.Connecting your radio

To connect your radio to your computer in order to fly the simulated models in Phoenix, you should first connect the USB connector end of your Phoenix USB interface to a spare USB 1. 1/2. 0/3. 0 port (see Fig. 4 - 2 - 6). You may hear a tone when you do so, and the first time you do this you may be informed that Windows is installing drivers. This will take a few seconds, then you will be informed that the device you have connected is ready to use.

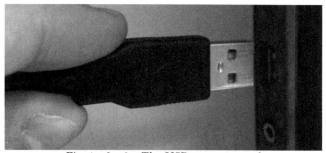

Fig. 4 - 2 - 6 The USB connector end

Once your USB interface is connected and has been detected by Windows, you will need to connect it to your radio's training port. If your radio requires an additional adaptor, please connect it now to the Stereo plug end of your Phoenix USB interface.

Now connect this to the training port of your radio unit (see Fig. 4 - 2 - 7). Most modern radios have a special "Simulator mode" which is activated by connecting the USB interface to the training port with the radio's power switch in the OFF position. If this is the case then when you connect your USB interface to your radio with the radio turned off, it will "power-on" automatically in this mode. This mode is very useful as all R/F signals are disabled which drastically improves battery life when using your radio with the simulator.

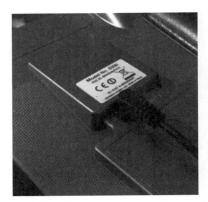

Fig. 4 - 2 - 7   The training port of transmitter

If your radio does not automatically "power on" when you do this, and you are sure that all connections are firm, and you have the correct adaptor fitted, then you may need to also switch your radio on at this stage.

2. Start Screen

By default, the Start Screen is shown when you first start Phoenix (see Fig. 4 - 2 - 8). This is a special menu which gives you instant access to any function or feature of Phoenix.

The Start Screen is composed of a number of customizable tiles, each which can be set to activate a menu item, change to a specific model or flying site, or load a previously saved Scenario.

To show or hide the Start Screen, press the TAB key on your keyboard at any time, or click the red "X" button in the top-right of the display when visible.

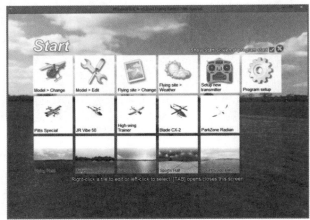

Fig. 4 - 2 - 8   Start Screen

3. Setup new transmitter

Clicking this menu item will launch the Setup new transmitter wizard (see Fig. 4 - 2 - 9).

This is a simple-to-follow wizard which will take you through all the steps needed to add a new radio to Phoenix, including calibration and control configuration.

This is a two-stage process starting with calibration and ending with selecting a control profile. You can fine-tune these settings later using the your transmitter and your controls menus.

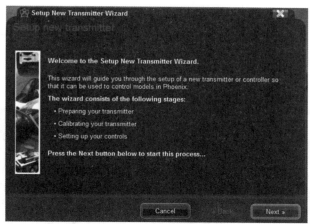

Fig. 4 - 2 - 9　Setup new transmitter wizard

(1)Calibrating your radio.

Clicking this will open the your transmitter menu (see Fig. 4 - 2 - 10), which contains functions for calibrating your radio for use with Phoenix.

The menu display shows a series of bars which each represent a channel coming from your radio.

When calibrated correctly, you should see each stick on your radio move one (or more)of the bars smoothly from one extent to the other, and centered when the corresponding stick is in its centre of travel.

Clicking the "Calibrate" button will start the calibration wizard.

You must do this each time you change radios, or make any changes in your radio unit.

Fig. 4 - 2 - 10　Transmitter menu

Before you can use your radio with Phoenix, it must be calibrated correctly so that the software knows the full range of movement of each stick on your transmitter, as well as the function of every programmable switch, knob or slider.

Click "Calibrate" to start the radio calibration wizard (see Fig. 4 - 2 - 11), which will take you through the steps required to successfully calibrate a new radio ready for use with Phoenix.

Click "Finished" to exit this menu.

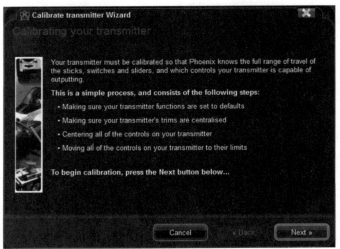

Fig. 4 - 2 - 11   Calibrate transmitter wizard

(2)Control-profiles.

This menu lets you select which channels coming from your radio will activate which controls on the simulated model.

For example, Channel 1 on your radio may be set to control the model's Throttle, and Channel 5 may control the model's retractable landing gear.

Phoenix saves these links between channels and model functions in a "Control Profile", and you can either select from a list of commonly-used profiles for various radio makes and models, or create your own custom profile entirely from scratch.

The list in Fig. 4 - 2 - 12 shows the current selection of saved control profiles, organized by radio manufacturer. You can expand a category by double-clicking the name of the category, or left-clicking the small arrow to the left of each category. Any custom control profiles you may create are saved in the "Custom" category at the top of the list.

To select a control profile, highlight it in the list with your mouse cursor, and then left-click to select it.

Click "New profile" button to start the New profile wizard which guides you through creating a new, custom control profile. You should use this if your radio model is not visible in the saved control profile list.

Fig. 4 – 2 – 12   Control-profiles

Click"Edit Profile" to edit the currently selected control profile. If you try to edit a preset profile, a copy of it will be made for you in the Custom category automatically.

(3)New profile wizard.

This wizard guides you through creating a new control profile for Phoenix.

You must have your radio calibrated successfully using the System > Your Transmitter > Calibrate wizard before you can use this menu correctly.

There are two choices when creating a new profile (see Fig. 4 – 2 – 13):

Quick Setup is a faster method which assigns the most common functions by detecting your stick movements.

Advanced Setup will create a new, blank profile for you and open it ready for editing.

Fig. 4 – 2 – 13   Create new profile wizard

(4)Edit profile menu.

This menu lets you assign any model function to any channel coming from your radio or any keyboard key (see Fig. 4 - 2 - 14).

Fig. 4 - 2 - 14　Edit control profile

The "Name of profile" lets you set the visible name of this profile as shown in the saved control profiles list.

The Controls list shows you each model function, and what input is currently set to control it, as well as the ability to adjust the minimum and maximum travel of the control, whether you wish to reverse the input or set an exponential curve for that input.

The left-hand label shows the model function. The Min edit-box lets you set the minimum travel for this function (0 - 100). The Status bar in the centre shows you the current state of this function. If set to a radio channel, this will change with your radio's stick inputs. The Max edit-box lets you set the maximum travel for the function (0 - 100). The Invert check-box lets you flip the input. The Curve button opens the Curve Editor where you can select from a range of different preset curves, or create your own.

The input drop-down box lets you select which input should control this function:

• Unmapped means the function is disabled.

• Controller channel 1 - 8 will mean that channel (radio stick, switch, slider or knob)will control the function.

• Static 0%/25%/50%/75%/100% lets you set a fixed value for this input.

When the job above is done, you can select Finished/Cancel button, which lets you return to the System > Your Controls menu with or without saving your changes.

4. Select the model

The Model Menu contains all model-specific functions, features and settings, including model selection and editing (see Fig. 4 - 2 - 15).

You can also use this menu to adjust settings for where your model will start on the field, and how it will be launched.

You can also easily find the most recently used models here and select them quickly.

Fig. 4 - 2 - 15　Model menu

Click "Change" to open the Change Model menu. This menu displays every model that you currently have installed on your simulator (see Fig. 4 - 2 - 16). Models are divided into categories and can be sorted by several different methods to make locating and selecting the model you want to fly easy and fast. This menu also shows additional information about the selected model and lets you select a variant or alternate colour scheme (if available).

Fig. 4 - 2 - 16　Select model

To select a model, highlight it in the list with your mouse cursor, and then left-click to select it. When you do so, the selected model will appear in the model preview pane to the right.

5. Launch

This menu contains options for setting how the selected model is launched on the field (see Fig. 4 - 2 - 17). The meanings of different launch options are as follows:

Automatic: Selects the most appropriate launch method for the selected model based on the model's type, power and undercarriage.

Ground launch: Always launch this model from the ground, sitting in the correct position on the field.

Hand launch: Throw this model using the mouse or your radio to set the throw direction, then use the mouse-button or throttle stick to launch.

Tow launch: For gliders, create a tow-plane which will drag your model into the air. Use the throttle stick to detach from the tow-line.

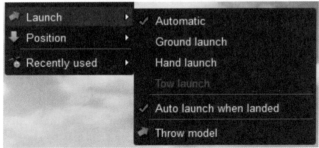

Fig. 4 - 2 - 17　Launch options

6. Select the flying site menu

The flying site menu contains all of the functions and features needed for setting up the environment you want to fly on, including the current field, weather conditions and additional objects in the scene (see Fig. 4 - 2 - 18).

You can also setup a layout (flags and markings) on the site from this menu as well as configure AI pilots to fly alongside you.

You can also quickly choose from the most recent sites flown on.

Fig. 4 - 2 - 18　Flying site menu

The "Change Flying Site" menu lets you select from a huge range of different locations to fly from. Phoenix comes with a great selection of gorgeous photographic panoramic sites, as well as our unique fully-3D "InfinityScape" sites where you generate your own infinite 3D landscape to fly across.

Flying site list shows the selection of sites you currently have installed on your system, as well as any 3D InfinityScape sites that have been previously created and saved (see Fig. 4 - 2 - 19). To select a site from the list, highlight it with your mouse-cursor and left-click the item. The preview pane will then change to show your new selection.

At the top of the site list are the 3D "InfinityScape" flying sites that you have previously created and saved. InfinityScape sites are full-3D landscapes that are generated from a selection of options that you can change, each resulting in a completely new, limitless terrain to fly across that has no map boundaries.

Fig. 4 – 2 – 19　Select the flying site

4.2.5　Training

1.Helicopter training

(1)Hover training.

This mode lets you more easily and quickly learn how to hover your helicopter by only requiring you to take control of one model function at a time while keeping the model at a set height (see Fig. 4 – 2 – 20). You can then gradually add more functions until you are controlling all of them confidently.

Please note that only helicopter models can be used with this trainer.

Fig. 4 – 2 – 20　Hover trainer

Setup:Use this drop-down box to select which function(s)you want to take control of.

Reset:Click this button to reset the model.

Enable/disable auto-restart:This option sets whether the model will automatically be reset when you move too far from the starting point, or the model tips too far over from level flight.

Normal/Inverted:This option sets whether you wish to practice normal or inverted (upside-down)hovering.

Orientation:Use this drop-down box to set the starting orientation of the model. You should be able to hover in all orientations confidently.

(2)Autorotation training.

This mode sets your model up for an autorotation from a preset height and speed and automatically cutting the throttle, letting you practice the manoeuvre without needing to fly the model to a starting position each time and saving you valuable training time (see Fig. 4 - 2 - 21).

Please note that only helicopter models can be used with this trainer.

Fig. 4 - 2 - 21   Autorotation trainer

Height:Select from a range of different starting heights for the manoeuvre.

Reset:Click this button to reset the model.

Enable/disable auto-restart:This option sets whether the model will automatically be reset after a landing.

Orientation:Select which direction the model starts from, as well as if you wish to attempt a normal or inverted autorotation.

2. Airplane training

(1)Torque training.

Similar to the hover trainer, this mode lets you practice torque-rolling or prop-hanging your model by only requiring you to take control of one function at a time (see Fig. 4 - 2 - 22). You can then gradually add more functions until you are controlling all of them confidently.

Please note that only airplane models can be used with this trainer.

Fig. 4 - 2 - 22   Torque trainer

Setup：Use this drop-down box to select which function(s)you want to take control of.

Reset：Click this button to reset the model.

Enable/disable auto-restart：This option sets whether the model will automatically be reset when you move too far from the starting point, or the model tips too far over from level flight.

Orientation：Use this drop-down box to set the starting orientation of the model. You should be able to prop-hang/torque-roll in all orientations confidently.

(2)Landing training.

This trainer sets your model up on approach to the landing position of the selected flying site, letting you practice the manoeuvre without needing to fly the model to a starting position each time and saving you valuable training time (see Fig. 4 – 2 – 23).

Please note that only airplane models can be used with this trainer.

Fig. 4 – 2 – 23　Landing trainer

Height：Select from a range of different starting heights for the manoeuvre.

Reset：Click this button to reset the model.

Enable/disable auto-restart：This option sets whether the model will automatically be reset after a landing.

Distance：Select how far the model starts away from your position.

(3)Thermal gliding.

In this single-player mode, you start high in the sky with a glider model and no throttle.

The aim is to ride the randomly created thermals and stay in the sky for as long as possible (see Fig. 4 – 2 – 24).

Only glider models can participate in this competition mode.

Fig. 4 – 2 – 24　Gliding training

4.2.6　Conclusions

Flying with simulator may be boring, but this is the essential path leading to veteran. You need more patience to adapt to this period and you will see the progress gradually, which may inspire you to go further.

Our teachers will give you all the information and skills you need to progress your career as a professional UAV pilot.

You should spend lots of time exercising nose—in hovering for rotorcraft and landing for airplane which are the core abilities of UAV flight, and try your best to do them well.

Words & Phrases

index finger ['ɪndeks] [fɪŋɡə(r)] *n.* 食指
ring finger ['rɪŋ] [fɪŋɡə(r)] *n.* 无名指
fingertip['fɪŋɡətɪp] *n.* 指尖
ascend[ə'send] *v.* 上升
descend[dɪ'send] *v.* 下降
joystick['dʒɔɪstɪk] *n.* 摇杆；控制杆
auxiliary [ɔːɡ'zɪliəri] *adj.* 辅助的
preliminary [prɪ'lɪmɪnəri] *adj.* 初步的；开始的
be applicable to [bi] [ə'plɪkəbl] [tu] 适用于
channel ['tʃænl] *n.* 通道
wizard ['wɪzəd] *n.* 向导
torque [tɔːk] *n.* 扭矩
prop-hang [prɒp hæŋ] 吊机

Exercises and Thinking

1. Translate the following sentences.

1) This method uses two thumb fingertips to move the left and right sticks, the index fingers and middle fingers are used to toggle switches, while the little fingers and ring fingers are responsible for holding the body of remote controller.

2) In order to improve the precision of flight operation, we need to proceed auxiliary exercise with remote controller that is exercising the precision of each channel under the supervision of "SERVO MONITOR" screen.

3) Similar to the Hover trainer, this mode lets you practice torque-rolling or prop-hanging your model by only requiring you to take control of one function at a time.

2. Answer the questions in your own words.

1) What is the most difficult during the flight simulation?

2) How to calibrate your radio with Phoenix R/C?

3) Please introduce the basic procedures of setting up new transmitter with Phoenix R/C.

Mission 3　Flying in the Sky

【Objective】 Enable your drone to take-off, fly and land safely without any damages.

【Analysis】 For flying, safety is the most important. You should keep the safety notes in mind and prepare for the flight such as charging drone battery, compass calibration check, locating landing zone and so on. And before formal flight, you need to be permitted to fly your drone by your teacher. If your assessment in mission 2 of this project is good or excellent, you can fly the drone with the training cable connected to your RC, which can get help from your teacher in emergency. After you behave well, you can fly alone.

Follow the instructions in this mission and obey the drone laws and regulations. Do exercises repeatedly until you are skilled in operating drones!

【Knowledge preparation】 Drone laws and regulations; flying principle.

4.3.1　Safety Notes

Fly only in safe areas, away from other people. Do not operate R/C (Remote Control)aircraft within the vicinity of homes or crowds of people. R/C aircraft are prone to accidents, failures, and crashes, due to a variety of reasons including lack of maintenance, pilot error, and radio interference. Pilots are responsible for their actions and damage or injury occurring during the operation or as of a result of R/C aircraft models.

Generally, R/C aircraft flies at high speed, thus posing a certain degree of potential danger. Choose a legal flying field consisting of flat, smooth ground without obstacles. Do not fly near buildings, high voltage cables, or trees to ensure the safety of yourself, others and your model. For the first practice, please choose a legal flying field and a training aircraft to fly for reducing the damage. Do not fly your model in inclement weather, such as rain, wind, snow or darkness.

R/C models are composed of many precise electrical components. It is critical to keep the model and associated equipment away from places that can cause contaminants. The exposure to water or moisture in any form can cause the model to malfunction resulting in loss of use, or a crash. Do not operate or expose to rain or moisture.

Before turning on your model and transmitter, check to make sure no one else is operating on the same frequency. Frequency interference can cause your model, or other models to crash. The guidance provided by an experienced pilot will be invaluable for the assembly, tuning, trimming, and actual first flight.

Operate this unit within your ability. Do not fly under fatigue and improper operation may cause danger.

During the operation of the aircraft, be conscious of your actions, and careful to keep your face, eyes, hands, and loose clothing away from the blades. Always fly the model a safe distance from yourself and others, as well as surrounding objects. Never take your eyes off the model or leave it unattended while it is turned on. Immediately turn off the model and transmitter when you have landed the model.

R/C models are made up of various forms of plastic. Plastic is very susceptible to damage or deformation due to extreme heat and cold climate. Make sure not to store the model near any source of heat such as an oven, or heater. It is best to store the model indoors, in a climate-controlled room temperature environment.

4.3.2 Safety Check Before Flying

Carefully inspect before real flight can reduce the aircraft crash accidents caused by different kinds of failures.

1.Environment safety check

The opening area with the minimum signal interference is the best choice for UAV flight which can lower and reduce the probability of accident.

(1)Before flight, the primary thing is observing the flying environment (see Fig. 4 - 3 - 1) to ensure there are no obstacles that influence the flight safety, such as electric wire, high voltage tower, base station and so on.

(2)Check if the flying zone is in No-fly zone and restricted area, please not to break any of the local laws and rules; In addition, the flying zone should be away from buildings and the crowds.

(3)Before flying, please check to make sure no one else is operating on the same frequency for the safety.

(4)Check if the weather is appropriate to UAV flight. Different types and designs of the drones can handle different types of severe weather. Check the drone's operating manual for specifications such as maximum wind speed and temperature limits. Commercial drones are usually designed to handle high winds and freezing temperatures. However, hobby drones are more susceptible to bad weather. You must also be especially careful with the wind, because even light winds or sudden gusts can cause you to lose control of your drone. Also, keep in mind that even returning to the fail-safe function may not save your drone in strong winds. Based on experience, it is best to fly only at wind speeds below 10 - 15 mile/h and avoid rain and snow. It is more likely that cloud cover will not affect drone connections, but they may be signs of rain. Pay close attention to the local weather forecast and land the drone as soon as it rains for the first time.

Fig. 4 – 3 – 1　Flying environment

2. Mechanical check

UAV is complicated and precise equipment, the body of UAV withstand great force, which may result in physical damages. Checking the body of UAV before flight can contribute to finding these damages in time to ensure the flight safety. The body check at least includes the items below:

(1)All UAVs are being inspected for possible cracking and corrosion. Check if your drone has any loose connections, cracks, recesses or any other inappropriate places.

(2)Before turn on the transmitter, please check if the throttle stick is in the lowest position and IDLE switch is OFF.

(3)Check for missing or loose screws and nuts. See if there is any cracked and incomplete assembly of parts.

(4)Carefully check main rotor blades and rotor holders for helicopter. Broken and premature failures of parts possibly result in a dangerous situation.

(5)Check the propellers for multicopter. Be especially careful when checking the propellers, as they are the key components responsible for the lift and thrust you need to fly. The important thing is that your propeller is free of debris, dents, cracks and dirt. Replace any damaged propellers before flying. Remember to perform this check not only at the beginning of the flight, but also between flights if you plan to perform multiple flights in a row.

(6)Check all ball links for helicopter to avoid excess play and replace as needed. Failure to do so will result in poor flight stability.

(7)Check the battery and power plug are fastened. Vibration and violent flight may cause the plug loose and result out of control.

(8)Check the connection between the drone and the controller and any other external devices (such as smart phones or computers).

3. The check and calibration of flight control system

Before each flight, you must check your equipment to make sure everything is in working order and ready to fly.

This procedure is the most complicated of the safety check before flight. Generally, it includes the following items:

First calibrate the compass. The compass is responsible for the flight direction of the drone. Correct calibration is essential for safe flight.

(1)Before operation, check every movement is smooth and directions are correct. Carefully inspect servos for interference and broken gear.

(2) Before flight, please check if the batteries of transmitter and receiver are enough for the flight. Make sure that the battery is fully charged. Check whether the battery fits tightly and will not loosen during flight. It is recommended to use a fully charged battery. A battery that is not used for a long time may not work well.

(3)Check all connected devices or payloads, such as sensors and cameras. Clean the camera lens and check to make sure your gimbal is in good working condition.

(4)Check if the number of satellites satisfies the demand of flying safety (see Fig. 4 - 3 - 2).

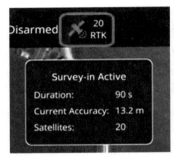

Fig. 4 - 3 - 2　The number of satellites

4. Power on/off the UAV

When turnon/off the unit, please follow the power on/off procedure. Power ON-Please turn on the transmitter first, and then turn on receiver, Power OFF-Please turn off the receiver first and then turn off the transmitter. Improper procedure may cause out of control, so please to have this correct habit. Check whether the drone is operating normally from hovering. Listen for unusual noises that may be signs of wear. Note any abnormal behavior that may indicate a connection or impending propeller failure.

Once you are sure that your drone is ready to fly, you can fly freely.

4.3.3　Real Flight

1. Fixed-wing plane
(1)Takeoff.

Takeoffs are relatively easy maneuvers to perform. You can make one takeoff with little effort and virtually no skill required. Nonetheless, takeoffs can be hazardous and have led to many crashes in real life.

As stated earlier, takeoffs should, if practical, be made into the wind. This procedure reduces both the ground speed and distance necessary to achieve flight, thereby minimizing wear and tear on the wheels and reducing the risk of a crash. You must select an appropriate runway, and then taxi the airplane to the departure runway.

Before actually taxiing onto the runway—and assuming that this is the first takeoff you have made during the flight—you should perform a pre-takeoff check. For example, when checking the left aileron, just turn the control stick counter-clockwise and ensure that the left aileron goes up and the right aileron goes down.

Likewise, check that the rudder moves left when the operator moves the rudder stick to the left on the transmitter and right when the rudder stick is in the right position. The trailing edge of the elevator should move up when the control stick is pulled aft and down when the control stick is moved forward.

In addition, you should adjust the attitude indicator, set the altimeter to the current altimeter setting, and set the heading indicator to the heading displayed on the GCS.

The takeoff involves three steps: takeoff roll, lift-off, and initial climb. Look down the runway to ensure that the path is clear and that no other aircraft are landing on the same or intersecting runways. I have delayed or aborted takeoffs when faced with other aircraft, teens on bicycles, or even a cow crossing the runway.

If the coast is clear, smoothly advance the throttle until it is fully open. After applying full power, check the engine instruments again to ensure that they continue to indicate normal operation (see Fig. 4 - 3 - 3).

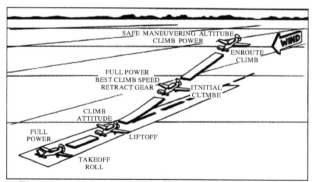

Fig. 4 - 3 - 3　Normal takeoffs are made into the wind

Because you have programmed out the left-turning tendencies, you do not experience the pull to the left during the takeoff roll, which requires the rudder to move right. Nonetheless, be careful to remain aligned with the runway and on or near the runway centerline. Should the airplane deviate to either side of the centerline, slight rudder deviation must be applied in the direction of the centerline.

The takeoff roll involves a transition from taxiing to flying. At the beginning of the takeoff roll, the fixed-wing drone steers with nose wheel steering and functions by having its weight on the wheels.

After takeoff, the ailerons, rudder, and elevator control the airplane. Somewhere during the takeoff roll, both systems play a role. For example, at 35 knots, both the nose wheel steering and the rudder play a role in steering the fixed-wing drone. As the airflow over the control surfaces increases, smaller amounts of control movement are required to affect the path of the fixed-wing drone.

As the airspeed passes through 50 knots, gradually pull the elevator stick to lift the nose wheel slightly off the runway. This process is referred to as rotation, and the speed at which it is initiated is called rotation speed. In strong, gusty winds, increase the rotation speed by one half of the gust factor to allow an added margin of safety. The main wheels quickly follow the nose wheel. It is necessary to carefully control the drone's attitude so that it does not settle back onto the runway or, conversely, so that it does not stall.

Allow the drone to accelerate to VY, the best rate of climb speed, which is approximately 85 knots indicated airspeed. Aileron control must be used to keep the wings level during the initial climb. Aircraft pitch must be adjusted to maintain the desired climb speed. With power fixed at full, speed adjustments are made with slight changes in aircraft pitch.

Upon reaching an altitude that is above any obstacles, accelerate the airplane to the enroute normal climb speed of 90 to 100 knots.

If a crosswind is present after takeoff, it will be necessary to make a wind correction. Turn the nose of the aircraft into the wind just enough to maintain a path aligned with the departure runway.

After passing through an altitude below the pattern altitude, make a medium banked turn to the left to fly out of the pattern 45° off the upwind track (on a heading of approximately 315 °).

(2)Landing.

Once established on final approach, extend the flaps to 30° and slow the airspeed gradually. Adjust pitch attitude as necessary to maintain the desired glide path to the touchdown point. The goal is to arrive at the runway with just enough airspeed to flare for landing and make a smooth touchdown.

Excess airspeed, as one would have if the approach angle were very steep, causes the airplane to float some distance down the runway. Where, as here, power is variable and available, pitch should be used for altitude control and power for airspeed. Nonetheless, coordination of the two is always necessary because of their aerodynamic relationship. The final approach involves a complex orchestration of many aerodynamic factors (pitch, power, flap setting, airspeed, and wind).

If it appears that the drone has dropped below an appropriate glide path, an increase in pitch (raising the nose) must be combined with an increase in power. If only the pitch is raised, the airplane will settle and touch down short of the touchdown target. If the drone is too high, the nose should be lowered and the power reduced so that speed does not increase above the target airspeed.

Full flaps should not be selected until you are confident that you could glide to the runway if the engine failed. The rate and angle of descent should constantly be reevaluated and adjusted so that the drone will arrive over the touchdown zone several decimeters above the runway. Power should be adjusted as necessary to maintain target speeds.

Moreover, the flap setting targets are only recommendations for a normal approach. They should not be mechanically applied. For example, if the drone is on a long final approach and appears to be low or slow, application of additional flaps should be delayed. It is inappropriate to retract flaps to correct for being low on the approach, because the retraction of the flaps will cause a sudden sinking of the airplane. A bad approach leads to a bad landing. If the approach looks bad, fly around the pattern and try again.

As the drone reaches the touchdown zone, adjust the pitch in a smooth, continuous motion to change from the approach attitude to the touchdown attitude. If power has not already been reduced to idle, it should be at this point. Without the flare, the drone flies into the runway nose wheel first—not a particularly good idea for impressing outside observers. Increase the pitch attitude at a rate that allows the airplane to gently settle to the runway while the excess speed (the speed above stall speed) bleeds off (see Fig. 4 – 3 – 4).

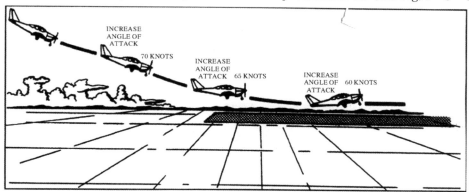

Fig. 4 – 3 – 4　During the flare, airspeed bleeds off while the pitch
attitude is raised to the proper touchdown attitude

The rate of pitch change during the flare is also a complex symphony of aerodynamic factors. If the flare is begun high, the rate of pitch change is less than if it is begun low. If the descent rate is high, the flare must be initiated earlier and may require a greater rate of pitch change.

Ideally, the drone's descent rate at touchdown is at or less than 3m per minute. The drone should initially contact the runway in a nose-high, tail-low attitude (see Fig. 4 - 3 - 5).

Fig. 4 - 3 - 5　A good flare attitude just prior to touchdown

After the main wheels contact the runway, pull the elevator stick to keep the nose wheel off the ground until the airplane decelerates and to take advantage of aerodynamic braking, which comes from the drag created by the high angle of attack. As the drone slows, push the elevator stick slightly to allow the nose wheel to lower to the runway surface. Directional control during the landing roll out is maintained with the rudder stick, which provide rudder control before the nose wheel touches the runway and nose wheel steering combined with rudder control after the nose wheel is down. Apply the brakes as necessary to bring the drone to a stop or to slow it to the pace of a walk before making a turn off of the runway.

(3)Traffic pattern.

The standard traffic pattern entry procedure usually involves a 45° entry onto the downwind leg. Fig. 4 - 3 - 6 depicts the standard traffic pattern (including pattern entry).

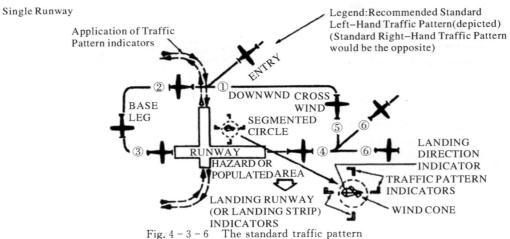

Fig. 4 - 3 - 6　The standard traffic pattern

The following steps provide the procedure for entering the standard pattern for arriving drone:

Enter pattern in level flight abeam the midpoint of the runway at pattern altitude.

Maintain pattern altitude until turning onto the base leg. Note that the descent should begin on the downwind leg after passing abeam the landing threshold.

Complete the turn onto the final approach at least one quarter mile from the runway.

The following steps are for departing drone:

Continue straight ahead until beyond the departure end of the runway.

If remaining in the traffic pattern, commence the turn to the crosswind leg beyond the departure end of the runway and within the reaching pattern altitude.

If departing the traffic pattern, continue straight out on upwind leg or exit on a 45° left turn beyond the departure end of the runway and after reaching pattern altitude.

Follow these steps:

When the airplane passes through 280m, commence a left turn onto the crosswind leg while continuing the climb to 300m.

As the aircraft reaches 300m, lower the pitch as necessary to maintain altitude. Fly the downwind leg approximately 200m from the landing runway.

Make a 90° left turn onto downwind at an appropriate time.

Reduce the throttle setting so that the airspeed does not exceed 130 knots.

When the drone passes abeam the midpoint of the landing runway, perform the Before Landing checklist. Turn the carburetor heat on before the throttle setting is further reduced.

It's a good idea to have target speeds for various points in the traffic pattern. For example, consider slowing to 90 knots on downwind after extending 10° of flaps, 80 knots on base after extending 20° of flaps, and 70 knots on final with full flaps.

Drop the first 10° of flaps and adjust the throttle to slow to 90 knots.

The turn onto base leg should begin when the drone is situated 45° off the approach end of the runway.

On the base leg, the drone must be descended from pattern altitude at a rate that allows it to arrive at runway elevation at the completion of the final approach.

Select a point near the center of the first one-third of the landing runway and target it as the touchdown point.

Extend the flaps to 20° during the base leg and, as discussed previously, the target speed on base leg should be 80 knots. If there is a wind blowing down the runway, there likely will be a crosswind while flying the base leg, which requires a crab angle to the left while flying a left pattern (see Fig. 4 – 3 – 7).

On all legs of the traffic pattern, an awareness of the wind and its effect must be maintained. Each leg should be flown on a heading that keeps the drone parallel to (on downwind and upwind)or perpendicular to (on cross-wind and base)the landing runway.

Make the turn onto the final approach at a safe altitude relative to terrain and obstructions and lead it so that the airplane rolls wings-level aligned with the runway's centerline. Many students have difficulty judging how much lead distance to allow for this, and this problem is aggravated by a tailwind on base leg or when transitioning to faster aircraft, which require more lead distance. Don't be disappointed if you overshoot or undershoot the turn to final several times. With practice, you will improve.

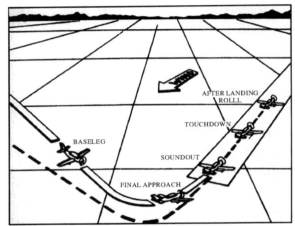

Fig. 4 - 3 - 7   A crab angle is often required while flying the base leg

Make the turn from base to final with a shallow to medium banked turn, because the drone is now slowed so that a steep bank could increase the stall speed to the speed at which the drone is flying.

Fig. 4 - 3 - 8 graphically depicts the increase in load factor (G units, which represent the force of gravity) as bank angle is increased while maintaining altitude.

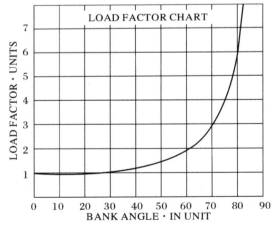

Fig. 4 - 3 - 8   Load factor increases dramatically at bank angles greater than 60°

For example, at 60° of bank in level flight, the load factor is 2 Gs, meaning that the effective load that the drone is supporting is twice its weight at rest on the ground. The drone's stall speed increases in proportion to the square root of the load factor.

What all this means is that using a steep turn from base to final could put you dangerously close to a stall. If, during the turn to final, you recognize that you will overshoot, do not steepen the bank beyond 35 or 40 degrees. Instead, make an s-turn to become realigned with the landing runway or abandon the approach and start again.

After completing the base to final turn, align the drone with the runway centerline. This is done to determine whether there is a crosswind. If there is, the drone will drift left or right of centerline. Assuming no crosswind, the drone should remain aligned with the centerline.

2. Rotor

(1) Tail-in hovering.

Take-off to hover, hold for five seconds. The tail of the UAV faces the pilot, try to keep it at a fixed point. This is the most basic subject. Most of the pilots start UAV flight from this item.

With the tail of the UAV facing towards you, you can control the aircraft in the most intuitive way, reducing the difficulty of control due to the visual orientation. Tail-In Hovering can exercise the pilot's basic reflexes in the initial control, and makes the pilot be familiar with the aircraft's control in pitch, roll, yaw and throttle. Completing the tail-in hovering practice means that the pilot enters the stage of "starting to play" from "cannot play".

Essentials: Please try to keep hovering at a fixed point with drifting slightly.

Cultivate the ability to correct the aircraft when there is a tendency to drift, which is essential for subsequent flights.

Don't be blindly self-satisfied, if you think that it's a success that you can control the UAV without crash, and let the plane drift around without correcting in time, it will cause greater difficulties in future flights.

Although boring, it is very important for you to fly well in this stage. If you feel that you have passed the level, then try again under wind level 5.

(2) Full lateral view hovering.

After the UAV takes off, make it hover at a fixed-point with nose towards left or right relative to the pilot. This is the first subject to break through after tail-in hovering. This kind of hovering can greatly enhance the pilot's sense of judging the attitude of the aircraft, especially the sense of distance. For a novice, it is very risky to practice this hovering directly, because the lateral tilt of the aircraft is difficult to judge. You can start from practicing 3/4 Rear View Hovering, so that you can use the conditioned reflex inherited from the tail-in hovering in the sense of orientation. When the 3/4 Rear View Hovering is OK, you can gradually practice Full Lateral View Hovering, and you will find it easier to complete.

It should be pointed out that most people have a lateral position (left or right) that they are accustomed to. This is normal, but don't just fly in one side position you are used to. You must practice both the left and right-side positions until you feel familiar with both sides.

Full Lateral View hovering is more difficult than tail-in hovering. It can be considered that hovering at a fixed-point for more than 7 seconds in a ball space with a diameter of 3 meters under wind grade 4 is a pass.

After completing Full Lateral View hovering, it means that small route flight is possible, and the controller can finally break through the boring hovering flight and start the route flight.

(3)Nose-in Hovering.

Take-off from the landing area (1 meter circle)to eye-level, hold momentarily.

Either Pirouette or Fly and stop in the Nose-in condition, the preferred method is to Pirouette to nose-in.

Hold in the Stationary Nose-in Hover for one minute. The Stationary Hover should give the appearance of being under total control.

Land with the skids completely within the landing area.

Although it is theoretically possible to fly on a small course after completing the Full Lateral View hovering, it is still recommended to continue to practice Nose-in Hovering.

For novices, Nose-in Hovering is extremely difficult, because in addition to the throttle, the attitude control is opposite to the Tail-in Hovering, especially for the pitch control, Pushing the forward-backward stick makes the UAV fly towards you, while pulling makes the UAV fly away from you. If a novice is not used to making mistakes, it is very dangerous.

You can try 3/4 Rear View Hovering, and then gradually turn to Nose-in Hovering, so that you can slowly adapt to the feeling of the control, which can effectively reduce the probability of crashing.

Nose-in Hovering is very important for route flight. You must cultivate the feeling of manipulation reflex. It is also quite good for pirouette practice in the future.

Flying UAV with the nose towards can bring you a wonderful feeling, just like the UAV is communicating face to face with the pilot.

The pass criteria for Nose-in Hovering is same as Tail-in Hovering. Try to control the aircraft in a 2-meter-diameter ball space for more than 10 seconds under wind grade 5.

(4)Small route flight.

After the UAV takes off, use the rudder to turn clockwise/counterclockwise to complete a closed sports field route. Small route flight is the first subject to be carried out after passing 4-position hovering criteria, it is the basis of all route flights.

For a pilot who is skilled in hovering with 4 positions (tail, two sides, and nose), he will find that it is quite simple to fly on small routes. Conversely, if you cannot pass the 4-position hovering, then even a small route flight is a challenge.

The trick to flying on small routes at the beginning is that you must pay attention to controlling the forward speed of the aircraft. High forward speed will bring unexpected difficulties for novices flying on small routes.

When turning, you should control the appropriate steering speed, do not hurry to turn around. When the 4-position hover is already proficient, a slow and rhythmic steering is the correct approach.

I have to emphasize again that both small clockwise and counterclockwise routes must be proficient in flying. Although it is more accustomed to most people to fly in one direction, a two-way proficiency route is essential for other subjects. If you don't practice well, you will encounter frustration sooner or later.

Please follow the high-standard specifications for small route flight practice. Don't be tempted to fly everywhere and let the plane go wherever you want. A pilot cannot just fly UAV casually which will bring no improvement in control, let alone the flying art.

The pass standard for the small route flight is: control the straightness of the route when flying in a straight line, and control the consistency of the left and right turn radius when flying in a turn. The speed and altitude should be as consistent as possible during the entire flight course. Achieve the above standards within the wind grade 4.

(5) Figure of eight—small route flight.

With the helicopter/multirotor flying straight and level after it passes the pilot, make a 270 degree turn away from the pilot, the aircraft will now be pointed directly at the pilot.

After the helicopter/multirotor is pointing at the pilot, execute a 360 degree turn, in the opposite direction. The aircraft will again be pointing directly at the pilot.

After the helicopter/multirotor is pointing at the pilot again, execute a 90 degree turn, in the same direction as the first 270 degree turn.

This maneuver must be done starting from both left to right (first 270 degree turn to the left, counterclockwise) and right to left (first 270 degree turn to the right, clockwise).

After the UAV takes off, use the rudder to turn to complete a figure of eight course clockwise/counterclockwise in the horizontal direction.

The figure-of-eight small route flight can help the pilot to become more familiar with the dimensional orientation and sense of controllability, which is very important for a comprehensive pilot.

If you are skilled in flying small route flight both clockwise and counterclockwise, then flying small route flight of figure-eight will be a piece of cake for you. If in actual flight, you still feel that it is more difficult to fly on the small route of figure-eight, it means that you have not really passed the small route flight or even 4-position hovering.

The figure-eight small route flight can to a large extent cultivate the pilot's adaptability to the helicopter position in the route, and can also practice the left turn and the right turn in a route at the same time. It is a must-practice subject for the primary route flight.

At first, you can choose one of the turning directions according to your own habits, but in the end, you must practice all of them, that is, turning clockwise on the left and turning counterclockwise on the right, or turning counterclockwise on the left and turning clockwise on the right.

The trick of the figure-eight flight is to control the speed of the aircraft according to your own ability, and constantly correct the attitude and orientation during the flight, and strive to achieve graceful and standardized movements.

The standard of figure-of-eight small route flight is: the left and right circles have the same flying radius, the figure-of-eight intersection is directly in front of the pilot, and the flight altitude and speed are the same throughout the route.

If you can basically meet the above standards under wind grade 4, it means you have passed the figure-of-eight small route flight.

When slow flight can no longer challenge your control ability, try to speed up the flight.

(6)Figure of eight—great route flight.

After the UAV takes off, it flies at a high speed and completes a figure-eight route horizontally.

The figure-of-eight great route flight is a step to make route flight more proficient, and is used to cultivate the pilot's ability to control the route flight in any direction.

After you have passed the great route of figure-eight, your flight proficiency can be greatly improved.

The trick to flying this course is to fly the great route clockwise and counterclockwise, and then control the flight speed and maintain a safe altitude. After a few flight attempts, gradually lower the altitude and increase the forward speed.

If you are pursuing flight art, you should also pay attention to the manipulation quality in daily flight. Try to maintain the same speed, height, left and right turning radius, and turning slope of the figure-of-eight flight route, and place the figure-eight intersection directly in front of the pilot.

The ability to achieve the above standards under winds of grade 4 to 5 indicates that you have passed the test of great route flight of figure of eight.

4.3.4　Conclusions

Remember the safety notes and do the flying preparations carefully. Pay more attention to the landing and nose-in hovering exercises, and try your best to have good performance.

Flying is no trivial, you should follow the instructions strictly. Only you are skilled in flying with simulator, can you do this mission. To get the most training from this mission, you need patience and diligence for maximum effectiveness.

Words & Phrases

vicinity [vəˈsɪnəti] n. 周围地区；附近
inclement weather [ɪnˈklemənt] [ˈweðə(r)] 恶劣的天气
contaminant [kənˈtæmɪnənt] n. 污染物
moisture[ˈmɔɪstʃə(r)] n. 潮湿；水汽
trim [trɪm] n. 微调

fatigue [fəˈtiːg] *n.* 疲劳

plastic [ˈplæstɪk] *n.* 塑料

deformation [ˌdiːfɔːˈmeɪʃn] *n.* 变形

debris [ˈdebriː] *n.* 碎片

dents [dents] *n.* 凹坑

crack [kræk] *n.* 裂纹

lens [lenz] *n.* 镜片

gimbal [ˈdʒimbəl] *n.* 云台

altimeter [ˈæltɪmiːtə(r)] *n.* 高度表

intersecting runway [ˌɪntəˈsektɪŋ] [ˈrʌnweɪ] 交叉跑道

airflow [ˈeəfləʊ] *n.* 气流

crosswind [ˈkrɒswɪnd] *n.* 侧风

bleed off [bliːd] [ɒf] 减小

touchdown [ˈtʌtʃdaʊn] *n.* 着陆

flare attitude [fleə(r)] [ˈætɪtjuːd] 拉平姿态

angle of attack [ˈæŋgl] [ɒv] [əˈtæk] 攻角；迎角

traffic pattern [ˈtræfɪk] [ˈpætn] 起落航线

crosswind leg [ˈkrɒswɪnd] [leg] 二边

commence [kəˈmens] *v.* 开始；着手

downwind leg [ˌdaʊnˈwɪnd] [leg] 三边

carburetor [ˈkaːbjʊrɛtə] *n.* 汽化器

obstruction [əbˈstrʌkʃn] *n.* 障碍

tailwind [ˈteɪlwɪnd] *n.* 顺风

base leg [beɪs] [leg] 四边

pirouette [ˌpɪruˈet] *n.* 快速旋转；定点旋转

tail-in hovering [teɪl ɪn] [ˈhɒvərɪŋ] 对尾悬停

3/4 rear view [rɪə(r)] [vjuː] hovering 45°角停悬

full lateral [ˈlætərəl] view hovering 侧面停悬

nose-in [nəʊz] [ɪn] hovering 对头悬停

figure [ˈfɪgə(r)] -of- eight -flying 水平 8 字飞行

Exercises and Thinking

1. Translating the following sentences.

1）R/C aircraft are prone to accidents, failures, and crashes, due to a variety of reasons including lack of maintenance, pilot error, and radio interference.

2）Make the turn onto the final approach at a safe altitude relative to terrain and obstructions and lead it so that the airplane rolls wings-level aligned with the runway's centerline.

3）Try to maintain the same speed, height, left and right turning radius, and turning slope of the figure-of-eight flight route, and place the figure-eight intersection directly in front of the pilot.

2. Answer the questions in your own words.

1) How do you overcome the difficulties during the flight training?

2) Please self-evaluate your performance in this mission.

3) List the safety notes that need to be paid attention to while flying your UAV.

【Project Evaluation】

Work activities		Weightage	Score
Reading and translation.		4%	
Preparation and Clearance		2%	
The flight dynamics of UAV		4%	
Flying with simulator	Simulator configuration	2%	
	Tail-in hovering	2%	
	Nose-in hovering	5%	
	Full Lateral View Hovering	3%	
	Autorotation	5%	
	Landing	8%	
	Bank-to-turn	3%	
Real Flight	Preflight preparation	2%	
	Safety check	3%	
	Tail-in hovering	2%	
	Nose-in hovering	6%	
	Full Lateral View Hovering	3%	
	Autorotation	7%	
	Figure of Eight	9%	
	Takeoff	3%	
	Landing	14%	
	Traffic Pattern	9%	
Attitude and attendance		4%	

Grade: □ Excellence □ Good marks □ Medium level □ Pass

【Project Conclusions】 This project aims to teach you how to fly your drone well. Firstly, You need to master the flight dynamics and know how the forces act on the drone firstly; then you can complete the flight training with simulator which help you to grasp the basic operations of your drone effectively; Finally, in order to adapt to the real flight and enhance your flight skill, you need to fulfil the real flight training.

Upon completion of this project, you can fly alone. Please conduct the safety check before flight and obey the drone laws and regulations.

Always keep in mind that safety is the first priority!

Project 5 How to Maintain and Repair a UAV?

【Description】 This project aims to teach you the basic UAV repair and maintenance methods. You need a laptop, telemetry radios, batteries, drone, consumable materials and maintenance tools. You are required to learn how to repair and maintain the UAV system, especially the analytical procedures and solution.

【Analysis】 As a kind of precise electronic and mechanical equipment, UAVs need to be operated correctly according to the standard and kept in good repair as recommended to ensure its normal flight and service life. UAV maintenance includes cleaning and drying, UAV battery maintenance, UAV motor maintenance and UAV GCS(Ground Control Station) maintenance, etc.

【Related knowledge and skills】 The structure and systems of UAV; RC technology; plant protection UAVs.

Read and translate some long and difficult sentences; know how to use the RC and GCS; be able to measure the current and voltage with the multimeter; know how to use oscilloscope.

Mission 1 Routine Maintenance of UAV Body

【Objective】 Students are required to discuss how to maintain the body of UAVs so that the UAVs can work normally.

【Analysis】 During the process of UAV operation , some faults may occur due to improper operation or lack of maintenance, affecting the normal use of UAV. Then it is necessary to analyze the possible causes of the faults and propose solutions.

【Knowledge preparation】 UAV structure; UAV navigation and communication technology; UAV assembly and commissioning.

In the process of operating a UAV, it may have the following faults.

5.1.1 Fault 1

Fault 1: The drone does not work after it's powered on. The LED indicator and flight controller indicator turn off, and the motors have no beep (self-check)sound.

The solutions are as follows:

(1)Check whether the output voltage of battery is above 22V;

(2)Check whether the plug is wrongly plugged in or not connected;

(3)Check whether there is abruption of power cord and plug;

(4)Open the airframe maintenance cover to check whether the flight control power supply indicator is normal;

(5)Check whether the airframe circuit board is damaged.

5.1.2 Fault 2

Fault 2:When the drone is powered on, the power plug sparks and rings out.

It may result from the following reasons:

(1)If the battery plug is directly connected to the drone without anti-ignition cable, it is normal for the power cord to strike fire (as the instantaneous current will become high when the high-voltage capacitor in ESC (Electronic Speed Control)is charged);

(2)The anti-ignition plug is damaged or the anti-ignition plug is not used correctly in order.

5.1.3 Fault 3

Fault 3:The drone cannot be armed for takeoff.

Causes of failure are listed as follows:

(1)The drone is not powered on normally;

(2)The calibration of remote controller is failed;

(3)GPS cannot work normally;

(4)There exists mismatch between RC and receiver.

The possible solutions are as follows:

(1)Check whether the drone is properly electrified.

(2)Check whether the user interface of the GCS for the RC push rod is in green state(the throttle is yellow). If it is red, click to collect the median position of the RC sticks in the setting interface. If it is invalid, click to calibrate the RC according to instructions after removing the propellers.

(3)Power off the drone and reconnect the battery.

(4)Check whether the data of drone is normal and whether the number of observable GPS satellites is 8 or above.

(5)Perform the combination stick command to arm the drone and execute "automatic loiter" and "RTL" command.

(6)Check whether the RC matches the receiver of the drone, if not (the receiver flashes in green or red), the receiver should re-link to the transmitter.

(7)Check whether the drone is in no-fly zone, where drone is prohibited for takeoff.

(8)Check whether there is any adjustment in the sub-trim of RC. If so, just return to "zero".

5. 1. 4 Fault 4

Fault 4: When the drone is placed on the ground and is in PosHold mode, you push the throttle stick, there are one or more of propellers that do not rotate.

The solutions are as follows:

(1)Place the drone on flat ground.

(2)Re-calibrate the ESCs.

(3)Check parameters to ensure we select the correct drone type(see Fig. 5 – 1 – 1) that matches the actual configuration, such as Octocopter X configuration or $+$ configuration, Hexacopter X configuration or $+$ configuration, etc. This setting will not be changed under normal circumstances. If changes happen, check whether the parameters are restored to the default.

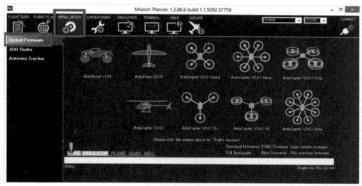

Fig. 5 – 1 – 1　Drone type

(4)Check whether the motor or ESC is damaged.

(5)Check whether the flight control output interface is damaged.

(6)Check whether the plug or connecting wire falls off, oris not firmly connected.

(7)Check whether the throttle stick is pushed for a long time. In general, the throttle stick is pushed to the middle or above position within 5 seconds (see Fig. 5 – 1 – 2).

Fig. 5 – 1 – 2　Push the throttle stick

5.1.5　Fault 5

Fault 5：Before taking off, the imbalance of the motor lasts more than 10 seconds. The solutions are as follows：

(1)Check whether the throttle stick is pushed overtime, in general, the throttle stick is pushed to the middle or above position within about 5 seconds, and the timespan should not be too long.

(2)Power off the vehicle and reconnect the battery.

5.1.6　Fault 6

Fault 6：After taking off, the drone drifts at fixed height or draws a large circle. The solutions are as follows：

(1)Recalibrate the compass.

(2)Check whether the drone is in "automatic loiter" mode.

(3)Check whether all RC sticks are in neutral position.

(4)Check whether the number of GPS satellites is more than 8. When the drone is flying around the high buildings or at the sheltered places, the GPS receiver will be directly impacted, so an open place is usually chosen as the take-off position.

(5)Check whether there is any imbalance of the motor lasting more than 10 seconds.

(6)Check whether there is any interference from high-voltage lines and high-power wireless equipment.

(7)Toggle back to manual mode, and do preflight check after landing.

5.1.7　Fault 7

Fault 7：The RC sticks do not work during flight or in loiter mode until we toggle to manual mode.

The solutions are as follows：

(1)Restart the GCS and drone (if the attitude mode control rod is pushed by mistake, there is warning that this mode is not allowed use without permission, or you will bear the consequences).

(2)Check whether the drone is in loiter mode. Sometimes the RC sticks do not work in mission mode.

5.1.8　Fault 8

Fault 8：The drone does not fly in a straight line during the flight.

The reasons and solutions are as follows：

(1)GPS positioning is inaccurate (high buildings in the neighborhood can block the signal reception, causing position errors).

(2)Check whether you push the rudder stick by mistake while you are pushing the elevator stick.

(3)Whether the GPS receiver is pointing straight ahead of the drone (recalibrate the compass after the adjustment).

(4)Check whether the direction of the drone is straightened, and there is no pushing sticks by mistake during flight.

5.1.9   Fault 9

Fault 9: The drone suddenly hovers uncontrollably during flight, then a few seconds later, it automatically rises about 20 meters, and flies towards the direction of the take-off point.

The solutions are as follows:

(1)Check whether the control stick is toggled to "RTL" mode. If so, toggle back to loiter mode.

(2)The RC antenna is damaged or blocked.

(3)There are high-voltage or high-power devices around with above ten thousand volts interfering with the radio signal transmission.

(4)Whether the receiver is damaged or not firmly connected.

(5)In RTL mode, it's normal for the drone to climb and then return to its home location. The take-off point is automatically recorded when GPS receiver observes above eight satellites each time. When the drone is in RTL mode, the drone will automatically rise to 20 meters high to avoid the obstacles if it is more than 25 meters far from the take-off point and then return to the take-off point. At the height of more than 20 meters, the drone returns to launch from its current height; when the height is below 20 meters, the drone will climb to the RTL height and then land.

The RC sticks are uncontrolled when the drone is in RTL mode. Only when it arrives at the point over the home location and begins to land, can the rudder and aileron sticks work. At the moment, the throttle stick is uncontrolled, so the operator can choose the flat place as the landing point. If you want to exit the RTL mode, you must toggle the "CH6" switch to loiter mode, and then toggle the "CH5" switch to manual mode. If the operator is not very skillful, the drone can be switched to manual mode first, then toggled back immediately to loiter mode, thus it can be controlled again.

Warning: When switching to the manual mode, don't pull the throttle stick to the bottom, and try to keep it at the neutral position, otherwise the drone will lose power and crash. In addition, pay attention to the battery voltage when returning to land. If the voltage of battery is lower than 21.9V, you must toggle to manual mode or loiter mode to land at its current position, avoiding crash due to running down of battery.

5.1.10 Fault 10

Fault 10:There is accidental vibration or continuous vibration during flight.

The operational steps to solve the problem are as follows:

(1)Check the parameter page of the GCS:whether there is low voltage alarm (the drone should land immediately when the voltage is below 21.9V, and it may crash at any time when the voltage is below 21.6V).

(2)Check the parameter page of the GCS:whether the difference value between built-in magnetic heading and external magnetic heading exceeds 10 (adjust the GPS receiver to be straight ahead of the vehicle and recalibrate the compass).

(3)Check the parameter page of the GCS:whether there is a fault down the page (the fault occurs when there exists external interference or improper operation, the fault can be resolved by cutting off the power of vehicle and restarting the vehicle).

5.1.11 Fault 11

Fault 11:There exits continuous motor imbalance or intermittent motor imbalance for more than 3 seconds during flight.

The solutions are as follows:

(1)Check whether the motor and propeller are on a parallel plane with the airframe, or simply confirm whether the tips of the two propellers are parallel (if not, loosen the arm screws and tighten them after leveling).

(2)If the motors continue to be unbalanced and their speed is still increasing, you should land the drone as soon as possible, check the ESCs and motors.

5.1.12 Fault 12

Fault 12:The drone is uncontrolled, and it vibrates back and forth towards one side.

The solutions are as follows:

(1)If you can skillfully fly the drone manually, you should toggle to manual mode to land the drone immediately.

(2)If you are not skilled in operating the drone manually, you should pull down the throttle stick as soon as possible to let the drone land and toggle back to manual mode to reduce unnecessary loss.

Words & Phrases

routine [ruːˈtiːn] *adj.* 常规的,日常的
abruption [əˈbrʌpʃən] *n.* 分裂,断裂
spark [spaːk] *n.* 火花,火星

vibrate [vaɪˈbreɪt] *v.* 振动

intermittent [ˌɪntəˈmɪtənt] *adj.* 间歇性的

drift [drɪft] *v.* 漂移

interface [ˈɪntəfeɪs] *n.* 接口

hexacopter [ˈheksəˌkɒptə(r)] *n.* 六旋翼无人机

octocopter [ˈɒktəˌkɒptər] *n.* 八旋翼无人机

PosHold mode n. 位置模式；定点模式

calibrate [ˈkælɪbreɪt] *v.* 校准

capacitor [kəˈpæsɪtə(r)] *n.* 电容

oscilloscope [əˈsɪləskəʊp] *n.* 示波器

multimeter [ˈmʌltɪmiːtə(r)] *n.* 万用表

Exercises and Thinking

1. Translate the following sentences.

1) Check whether the drone is in no—fly zone, where drone is prohibited for takeoff.

2) Check parameters to ensure we select the correct drone type that matches the actual configuration, such as octocopter x configuration or + configuration, hexacopter x configuration or + configuration, etc.

3) Only when it arrives at the point over the home location and begins to land, can the rudder and aileron sticks work.

2. Answer the questions in your own words.

1) What are the possible reasons of arming failure?

2) When the drone is in RTL mode, what will the done do?

3) What is the functions of GCS in this mission?

Mission 2 Routine Maintenance of UAV GCS

【Objective】 To be able to solve common GCS problems, such as the detention of communication and inconsistent information display , etc.

【Analysis】 When the GCS (see Fig. 5 − 2 − 1) is used for mission planning, there are often cases that the communication between the GCS and UAV is disconnected, or the UAV information displayed on the GCS is inconsistent with the actual situation. In the cases, troubleshooting is necessary, thus we can find out the causes and provide solutions to make the GCS work properly.

【Knowledge preparation】 UAV flying modes; UAV navigation and communication technology; electromagnetic interference.

In the process of operating UAVs, due to inadequate maintenance of the drone and GCS, the drone may have the following faults.

Fig. 5 - 2 - 1　UAV ground control station

5.2.1　Fault 1

Fault 1：GCS completely or intermittently disconnects communication with the drone.

The solutions are as follows：

(1)Check whether the wires of drone are connected properly.

(2)Check whether the battery level of the GCS is normal and whether the power switch is turned on correctly.

(3)Check whether the GCS radio antenna(see Fig. 5 - 2 - 2) and the radio telemetry antenna are firmly connected. If not，you should check the wiring connection and do the communication test to ensure the function is normal.

Fig. 5 - 2 - 2　GCS antenna

(4)Check whether smartphones or electronic devices around are connected to the WIFI.

(5)It is normal that the drone occasionally disconnect communication with GCS after take-off for no more than 3 seconds，which is related to the electromagnetic interference or weather.

(6)Check whether there are tall shielding objects or electromagnetic interference equipment around (such as high voltage lines，power plants，etc.)which will affect the normal transmission and reception of signals.

(7)Put the antenna as high as possible. If the antenna is omnidirectional，you can't make the antenna of GCS point towards the drone.

5.2.2　Fault 2

Fault 2：The GCS data display doesn't correspond with the actual status of the drone.

The causes of the fault and solutions are as follows：

(1)Fault：The altitude showed on the GCS is not the real altitude.

Solution：As the altitude of the drone is measured by the barometer，which is always closely related to the local weather at the time. So the barometer may be defective or there is function failure of the sponge that covered on the barometer.

(2)Fault：The voltage of drone in the air is different from that of drone on land.

Solution：It is a normal case that the battery voltage increases (no more than 1V) when the drone lands as the weight of load drops sharply after landing.

(3)Fault：There is a deviation between the actual throttle value and the RC throttle value.

Solution：The actual throttle value is corrected by the flight controller. If the data exceeds the threshold，check whether the drone is overloaded.

Words & Phrases

antenna [æn'tenə] *n.* 天线

electromagnetic [ɪˌlektrəʊmæg'netɪk] *adj.* 电磁的

shield [ʃiːld] *v.* 给……加防护罩

omnidirectional [ˌɒmnɪdə'rekʃənl] *adj.* 全向性的

sponge [spʌndʒ] *n.* 海绵

barometer [bə'rɒmɪtə(r)] *n.* 气压计

Exercises and Thinking

1. Translate the following sentences.

1) It is normal that the drone occasionally disconnect communication with GCS after take–off for no more than 3 seconds，which is related to the electromagnetic interference or weather.

2) There is a deviation between the actual throttle value and the RC throttle value.

3) If the data exceeds the threshold，check whether the drone is overloaded.

2. Answer the questions in your own words.

1) What are possible causes of the fault that GCS completely or intermittently disconnects communication with the drone?

2) What is the cause that the altitude showed on the GCS is not the real altitude?

3) What will the voltage of drone change when the payload of drone becomes lighter?

Mission 3　Routine Maintenance of Remote Controller

【Objective】To be able to analyze and find out the causes, then propose solutions to the faults of RC in the daily use.

【Analysis】Remote controllers often have some problems in use such as linkage failure, abnormal startup sound and the wrong corresponding action of the control sticks. To solve these problems, you are required to have a certain understanding of the working principle of RC, to master the communication mechanism and generative process of control commands.

【Knowledge preparation】The working principle of RC.

The maintenance of RC include cases as follows.

5.3.1　Fault 1

Fault 1: There is beep sound after turning on the RC.

The solutions are as follows:

(1) Check whether the throttle stick position of the RC is pulled down to the bottom.

(2) Check whether the voltage of the RC reaches the preset alarm value (standard initial setting is 6.8V).

5.3.2　Fault 2

Fault 2: The drone cannot react correctly to the control sticks of RC.

The solutions are as follows:

(1) check the channel setting of RC (e. g. The default factory setting of Japanese Hand: channels 1 – 6 are J1, J3, J2, J4, SE, SC; The default factory setting of American Hand: channels 1 – 6 are set as J1, J2, J3, J4, SE, SC).

(2) Check whether the signal wires between flight controller and RC receiver are correctly connected.

(3) Check and calibrate the RC (do not forget to remove all the propellers).

5.3.3　Fault 3

Fault 3: The receiver cannot be linked with the transmitter.

The solutions are as follows:

(1) Turn on the RC, then power on the drone. If receiver status indicator becomes blinking green which means the receiver is waiting for link, then press the reset button on the receiver until the receiver status indicator flashes a red light for three times. When the receiver status indicator becomes solid green, the link procedure is successful

(see Fig. 5 – 3 – 1). Lastly, power on the receiver again and check whether the receiver is being paired with the transmitter.

Fig. 5 – 3 – 1   Link mode

(2)Do not press thereset button of receiver for a long time to avoid changing the receiver mode.

(3)Check whether the RC control sticks and switches are in the initial positions before flight (the switches are placed in the OFF position, the throttle stick is put at the bottom).

Words & Phrases

mechanism ['mekənɪzəm] *n.* 机制
generative ['dʒenərətɪv] *adj.* 有生产力的；能生产的
preset [ˌpriː'set] *adj.* 预先调整的
solid green ['sɒlɪd] [griːn] 常绿

Exercises and Thinking

1. Translate the following sentences.

1) Remote controllers often have some problems in use such as linkage failure, abnormal startup sound and the wrong corresponding action of the control sticks.

2) There is beep sound after turning on the RC.

3) When the receiver status indicator becomes solid green, the link procedure is successful.

2. Answer the questions in your own words.

1) What does it mean when receiver status indicator becomes blinking green?

2) What are the initial positions of RC control sticks and switches before flight?

Mission 4   Routine Maintenance of Batteries

【Objective】 To be able to propose solutions to charge fail, battery swelling, and even explosion.

【Analysis】 LiPo batteries(see Fig. 5 – 4 – 1) offer an advanced charging and power storage solution for people looking to increase the life span of their charged batteries over traditional nickel-cadmium and nickel-metal hydride cells. LiPo batteries are often preferred in electric remote control models for their longer life. Unlike other rechargeable batteries, LiPo batteries present serious and imminent dangers if charged or discharged incorrectly.

Fig. 5 - 4 - 1　Lipo battery

Battery swelling / bloating is basically caused by too much heat, vast changes in temperature and/or moisture getting into the battery itself. The lifespan and condition of the battery will depend on the totality of how it was used and stored. In addition, every electronic product and batterie has a certain warranty period and lifespan. For safety purposes, please do not use bloated or bulging batteries when flying or operating your drone so that the drone's performance will not be affected or compromised. Using bloated or bulging batteries might lead to damages of the aircraft, or worse, damage to property. Battery being a bit old, there may be some gas in the battery that expanded on that high altitude and then got lesser swelling at lower altitude. It's like blowing up a baloon at sea level and the baloon becomes much bigger when it reaches high altitude.

Overcharging a LiPo battery will first damage the battery itself. A LiPo battery will swell noticeably if overcharged, rendering it useless because of damage caused by the swelling. The heat generated from an overcharged LiPo battery will also damage its charger and may cause harmonics to resonate through the electric circuit and damage other electronic devices in the home. If left unattended, an overcharged LiPo battery will continue to swell until it explodes. To prevent overcharging, these batteries should never be connected to any charger that isn't specifically designed for LiPo batteries. All safety instructions for the charger should be followed to the letter, and avoid any battery displaying signs of overcharging.

Ruptured LiPo batteries will ignite anything flammable nearby. When LiPo batteries are charged in the home, the flames from an exploding battery could cause curtains, furniture, carpet and other common household items to catch fire. If allowed to burn, the fire caused by this could consume the entire house or building in which the battery exploded, leading to catastrophic levels of fire damage, property loss and injury or death. As a safety precaution, LiPo batteries should never be charged indoors, and fast access to a fire extinguisher should be available.

When a LiPo battery explodes, superheated pieces of both the housing and internal components are torn from it and thrown into the air. Any shrapnel from the LiPo battery explosion will cause serious burns on the skin of anyone hit by it. The intense heat and trajectory of debris from a LiPo battery explosion can cause injuries by burning through clothing, or by igniting clothing which then burns the wearer over more extensive parts of the body. Anyone charging LiPo batteries should have easy access to a first aid kit and should remove any pieces of battery from her skin immediately on contact, then treat the wounds quickly to avoid more severe injuries.

【Knowledge preparation】 UAV flying modes; aeronautical meteorology.

Improper use of batteries may cause many faults and even bring property loss and personal injury. The causes and solutions are as follows.

5.4.1 Fault 1

Fault 1:The plug of battery is abnormally hot after flight and the battery cannot be charged.

Causes of faults:

(1)Improper welding of battery plug.

(2)Loose connection of battery plug.

(3)Short-circuit of battery balancing plug.

The solutions are as follows:

(1)Check whether the battery plug is improperly welded or broken, and whether the welding meets the technical requirements.

(2) Check whether the battery plug and the dronepower plug are not firmly connected, shorted, or incompatible (frequent plugging and unplugging will affect the connection resulting in loose connection). If so, the drone plug should be externally expanded to increase the contact area.

(3) Check whether the battery balancing plug is shorted or virtually connected, which can result in charge fail.

5.4.2　Fault 2

Fault 2:Battery is swelling.

Causes of faults:

(1)The user does not use the suited charger as required;

(2)Frequent deep charge and discharge cause battery swelling easily;

(3)Frequent overcharge, over-discharge and high charging current;

(4)Being used below 0 degree or in a high temperature for a long time;

(5)The battery is charged immediately when the temperature of battery is too high after use.

The solutions are as follows:

(1) Use the professional charger provided by the original manufacturer;

(2) Avoid frequent deep discharge, and keep reasonable charging and discharging;

(3) Avoid overcharge and over-discharge of batteries;

(4) Choose suitable flying environment to ensure the working efficiency of the drone;

(5) Place the used battery for more than five to ten minutes before charging.

5.4.3　Fault 3

Fault 3: Battery swells or explodes in transportation and storage.

Causes of faults:

(1) Improper protective measures are taken when the battery is in transportation.

(2) The battery is storaged with low voltage or high voltage for a long time.

(3) The battery is storaged in the humid and high temperature environment.

(4) No supervision on the battery charging site.

The solutions are as follows:

(1) Take protective measures when the battery is in transportation. Discharge the battery until it reaches the storage voltage; adopt anti-deformation protector as the outer packaging and soft protector as inner packaging.

(2) Keep the voltage between $3.8-3.85V$ (storage voltage) when the battery is not in use for more than 3 days. Too high voltage will be a risk of spontaneous combustion, while too lowvoltage will affect the lifespan, so a charging and discharging cycle is required every 3 months.

(3) Store the battery in the environment of temperature between 0 ℃ and 20 ℃. As the self-discharge of lithium battery is affected by ambient temperature and humidity, which will accelerate the self-discharge of the battery.

(4) Power off the charger after the battery is fully charged. The charging process shall be supervised all the time.

5.4.4　Fault 4

Fault 4: Battery cannot be charged normally.

The solutions are as follows:

(1) Check whether the battery balancing plug and the charger connection socket are firmly connected or not.

(2) Check whether the button of charger is pressed correctly after the battery is connected.

(3) If the charging time of the charger under AC 220V voltage does not match that of the generator in the workplace, check whether the power of the generator corresponds with that of the charging station.

Words & Phrases

swell [swel] v. 膨胀；鼓出

moisture ['mɔɪstʃə(r)] n. 潮湿

rupture ['rʌptʃə(r)] v. 断裂，裂开，破裂

welding ['weldɪŋ] n. 焊接

precaution [prɪ'kɔːʃn] n. 注意事项；预防措施

accelerate [ək'seləreɪt] v. 加速

combustion [kəm'bʌstʃən] n. 燃烧

spontaneous [spɒn'teɪniəs] adj. 自发的

aeronautical meteorology [ˌeərə'nɔːtɪkl] [ˌmiːtiə'rɒlədʒi] n. 航空气象学

Exercises and Thinking

1. Translate the following sentences.

1) Check whether the battery plug is improperly welded or broken, and whether the welding meets the technical requirements.

2) Choose suitable flying environment to ensure the working efficiency of the drone.

3) Take protective measures when the battery is in transportation.

2. Answer the questions in your own words.

1) What measures should be taken to prolong the lifespan of Lipo battery?

2) What measures should be taken to ensure the safe transportation of Lipo battery?

3) What are the possible causes of fault that battery cannot be charged normally?

Mission 5　Routine Maintenance of Spraying System

【Objective】To solve problems related to the spraying system of plant protection UAV, such as sprinkler leakage, function failure of pump or poor performance.

【Analysis】The spraying system mainly consists of spray tank, pump motor, hoses, sprinklers and nozzles. The pump motor pumps pesticide from spray tank into hose, then the pesticide will reach the sprinkler, and equal distribution of the liquid outflow from the spray nozzle forms a wide spectrum of drops. You need to have a fundamental understanding of the working principle of the spraying system, and learn from the user manual and safety guidelines provided by the manufacture.

【Knowledge preparation】The structure and principle of spraying system.

There are typical maintenance cases of the spraying system as follows.

5.5.1　Fault 1

Fault 1:The pump(see Fig. 5 - 5 - 1) doesn't work and there is no sound.

The solutions are as follows:

(1)Check and make sure the pump powerplug is firmly connected.

(2)Check whether you activate the function of switch that control the pump.

(3)Check whether the receiver is correctly connected to the spraying system and whether there exists virtual connection.

(4)Check whether the loose receiver is linked to the transmitter.

(5)It is necessary to power on or restart the RC after the radio link procedure.

(6)Check whether the buck module works normally.

(7)Check whether the ESC of the pump works normally and whether there is a voltage output.

(8)Check whether the pump works normally by performing a test. Pay attention to the positive and negative poles.

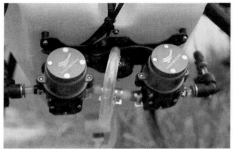

Fig. 5 - 5 - 1　The pump

5.5.2　Fault 2

Fault 2:The pump functions well but the spraying system cannot spray normally.

The solutions are as follows:

(1)Check whether there exists spatial interference between hoses, which may affect the flow of pesticide. In this case, dismount the hoses and re-mount them correctly.

(3)Check whether the inlet of the pump is connected to the spray tank, and whether the outlet is connected to the T-junction.

(4)Check whether the hoses are blocked and pay attention to cleaning the spray tank.

(5)If the pressure of pump is not enough to pump liquid, check whether there are holes on the lid of the spray tank, if not, punch holes to ensure the air circulation.

5.5.3 Fault 3

Fault 3:Nozzle leaking.

The solutions are as follows:

(1)Check whether the connection between hose and nozzle(see Fig. 5 - 5 - 2) is tight.

(2)Check whether the nozzle anti-drip valve (knob on the top nozzle)is loose and leaky, if so, tighten it.

(3)Check whether there is a rubber gasket inside the nozzle.

(4)It is recommended to clean the nozzle thoroughly after each spraying operation. When cleaning, pay attention to the mounting sequence of inner parts of the nozzle (purple nozzle-rubber gasket-yellow filter net). Don't forget any items (as the nozzle is a ceramic product, and it mustn't be disassembled roughly).

Fig. 5 – 5 – 2   The nozzle

5.5.4   Fault 4

Fault 4:The spraying performance is not good.

Make sure to calibrate the flow meter before your first operation. Otherwise, the spraying performance may be adversely affected.

1.Preparation before calibration:discharging the bubbles in the hoses

(1)Fill the spray tank with approximately 2L of water.

(2)Use the automatic bubbles discharge function to discharge the bubbles according to the descriptions in the Discharging the Bubbles in the Hoses section of instruction. You can also manually discharge the bubbles. Press the spray button to discharge the bubbles and press the button again once all bubbles are discharged.

2.Flow meter calibration

(1)In the app or GCS, tap Execute Operation to enter Operation View. Tap 🟡, then ⌖, and tap Calibration on the right of the flow meter section.

(2)Calibration starts automatically. After 25 seconds, the result of the calibration is displayed in the app or GCS.

• After calibrating successfully, you can proceed with the operation.

• If calibration fails, tap"?" to view and resolve the problem. Afterward, recalibrate.

3.When to recalibrate

(1)Installing a different nozzle model. Note:Choose the corresponding model in the app or GCS after replacing nozzles. Go to Operation View, tap 🟡, then ⌖ for configuration.

(2) Using a liquid of a different viscosity.

(3) The error between the actual value and the theoretical value of the completed area is more than 15%.

Words & Phrases

spray [spreɪ] *n.* 喷雾

hose [həʊz] *n.* 软管

sprinkler ['sprɪŋklə(r)] *n.* 喷洒器

nozzle ['nɒzl] *n.* 喷嘴

pump [pʌmp] *n.* 泵

discharge ['dɪstʃɑːdʒ] *n.* 放电

pesticide ['pestɪsaɪd] *n.* 杀虫剂,农药

rubber gasket ['rʌbə(r) 'gæskɪt] *n.* 橡胶垫片

Exercises and Thinking

1. Translate the following sentences.

1) The pump motor pumps pesticide from spray tank into hose, then the pesticide will reach the sprinkler, and equal distribution of the liquid outflow from the spray nozzle forms a wide spectrum of drops.

2) Check whether there exists spatial interference between hoses, which may affect the flow of pesticide.

3) It is recommended to clean the nozzle thoroughly after each spraying operation.

2. Answer the questions in your own words.

1) What is the function of spraying system?

2) Why does the nozzle leak?

3) What factors influence the spraying performance?

Mission 6　Unkeep of UAVs

【Objective】 Keep in good repair of the UAV main subsystems, such as body, battery, remote controller and so on.

【Analysis】 As a new high-tech product, UAV should be operated and used properly. In addition, its routine unkeep and check are also crucial. Generally, UAV pilots have a one-sided understanding that they know the design, assembly, debug and flying of UAY is enough. Moreover, because of few continuous flying missions, they do not pay much attention to maintenance. It is very common for the pilots to directly pack the UAV after flying and then take it out to fly directly next time. As the missions increase, problems gradually emerge. In order to prolong service life of UAV, it's necessary to do unkeep of UAY after a flight, especially the body, battery and RC, for they are the three main parts of the UAV system. For the body, its mechanical connections should mainly be checked; for the battery, charging & discharging and storage methods should be noticed; for the RC, we should concern its use environment.

【Knowledge preparation】UAV structure; the principle of Lipo battery; the working principle of RC.

5.6.1　Body Unkeep

(1)Clean the airframe and spraying system after a flight to reduce the body weight which ensure its endurance flight.

(2)Check the state of drone fasteners after a mission.

(3)Check whether there are aging, loose connection of power plug.

(4)Clean the motors and do the motor test to check whether there is abnormal noise of motor.

(5)Remove all the propellers, push the throttle stick to check whether the arms have abnormal vibration.

5.6.2　Battery maintenance

(1) Do not over-discharge: once single-chip cell voltage of the battery is below 3.65V(see Fig. 5 – 6 – 1), the voltage's drop speed will be accelerated and it will lead to over-discharge which will bring damage to battery (eg battery swell)and even may cause drone crash.

Fig. 5 – 6 – 1　DJI battery parameters setting

(2)Do not overcharge: use the appropriate charger to charge with the required charging current. It's forbidden to charge the battery with the current that exceeds the prescribed value and successive charging for a long time after the battery is fully charged is also not allowed.

(3)Do not store the fully charged battery: fully charged battery cannot be kept for more than 3 days. If the batteries are not discharged within a week, some batteries will directly swell, and some batteries may not swell temporarily, but they will be directly scrapped after being stored with fully charged state for several times. Therefore, the correct way is to charge the battery after receiving the flying mission. If there is no flying mission in 3 days after the battery is used or the fully charged battery hasn't been used in 3 days due to various reasons, please discharge the battery with single-cell voltage between 3.80V and 3.85V. If the battery is not in use within three months, the battery can be kept continuously after one charge – discharge, which can prolong the battery life. Battery should be stored in a cool and dry place. For long-term storage, the best way is to put the battery in a sealed bag or sealed anti-explosion tank, with recommended environmental temperature of $10 - 25℃$.

(4)Do not damage the outer casing: the battery outer casing is an important structure which can prevent the battery from explosion and catching fire resulting from liquid leakage. Damage to the aluminum plastic casing of lithium battery will directly lead to fire or explosion. The battery should be handled with care. When fixing the battery on the drone, the tie wrap should be tightly tied. Because it is possible that the battery will be thrown off due to untight tie wrap when the drone is conducting a large dynamic flight, which will easily cause damage to the battery outer casing.

(5)No short circuit: it often occurs in the process of battery bond wire maintenance or transportation. Short circuit can directly produce sparks or make the battery explode. When the battery power cord needs to be re-welded, it is particularly important to pay attention: the solder iron cannot touch the positive and negative poles of the battery at the same time. In addition, in the process of transporting batteries, the best way is to cover each battery with a self-sealing bag and place them in the explosion-proof tank to prevent them from short circuit springing from that battery's positive and negative poles touch any conductive object at the same time due to bumping and collision.

(6)Do not place batteries in low temperature environments: many pilots may ignore this point. As it is very cold in north of China or in high altitude areas, the battery's discharge performance will become worse if it is placed outside for a long time in such weather conditions. If you fly the drone with the same time as you fly under normal temperature, the drone may crash, you are recommended to set a higher alarm voltage (e. g. single-cell alarm voltage is adjusted to 3.8 V), as in low temperature environment the voltage descend very quickly. So the time of flight in low temperature is recommended as half of that in normal temperature and you should land the drone as soon as the drone alarms. Moreover, store the battery in a warm environment before a flight, such as in a house, a car or a heat preservation, etc and load the battery quickly before take-off.

5.6.3   Attentions in Use of RC

1. Use precautions

(1)Charge with the appropriate charger.

(2)Do not charge the battery when the RC is powered on, and do not debug the RC while charging.

(3)Keep RC far away from strong magnetic environment.

4)No strong vibration.

(5)It's necessary to do the dust and water prevention work regularly.

2. Safety precautions

(1)Be familiar with the flying environment, stay away from high obstacles and the crowd, and confirm there are no any unsafe factors.

(2)Keep the throttle stick of the RC at the lowest position before takeoff.

(3)Stay away from high humidity environment and do not fly in windy weather or on rainy days, which will cause the electronic equipment failure.

(4)Stay away from high temperature heat source, so as not to cause damage to electronic equipment or other components.

(5)It is suggested to do flying test of the drone without propellers to check if the RC and motors can work normally. Only if everything is OK, can you install all propellers.

(6)Prepare rescue tools, such as smartphones and walkie-talkies.

(7)Fly under safe take-off weight, and do not overload the drone.

(8)Before flight, check whether the UAV system is in normal status and whether there is co-channel interference.

(9)Keep away from working drone. Do not touch the rotating motor or propellers and even grasp them with your hands.

(10)Do not wear loose-fitting clothing.

Words & Phrases

unkeep [ʌnˈkiːp] *n.* 检修；维护

scrap [skræp] *v.* 报废

throw off [θrəʊ ɒf] 摆脱，甩掉

bond wire [bɒnd] [ˈwaɪə(r)] 焊接线

solder iron [ˈsəʊldə(r)] [ˈaɪən] 电烙铁

loose - fitting [ˌluːs ˈfɪtɪŋ] adj. 宽松的

Exercises and Thinking

1. Translate the following sentences.

1) If the battery plug is directly connected to the drone without anti-ignition cable, it is normal for the power cord to strike fire.

2) When the drone is flying around the high buildings or in the sheltered places, its GPS will be directly impacted, so an open place is usually chosen as the take-off ground.

3) Make use of the professional charger provided by the original factory, as the unqualified charger may easily cause expansion of the battery.

2. Fill in the blanks with proper form of the given words.

1) The quicker it _____ (vibration), the greater the frequency of sound it produces.

2) If you _____ (maintenance) a road, building, vehicle, or machine, you keep it in good condition by regularly checking it and repairing it when necessary.

3) If the battery balancing plug is shorted or loosely _____ (connect), it will result in abnormal charging.

4) Press the button on the _____ (receive) until the red light is flashing for three times and then release it.

5) Hundreds of lives could be saved if the _____ (install) of alarms was more widespread.

6) The _____ (full) _____ (charge) battery can make the drone run for a long time.

7) Do not touch the _____ (rotate) motor or propellers with your body, and do not grasp it with your hands.

8) Today, international exchange is getting _____ (frequency) day by day, English has already become a basic skill.

9) The smartphones or electronic devices around will _____ (interference) with the radio communication at the airport.

10) Restore factory default parameters of the RC, and do not change them without authorization except _____ (profession).

3. Read the following passage and answer questions.

How to Maintain and Take Care of Your Drone?

A pre-fight check is always important when it comes to flying your drone. A pre-fight check is basically a list of tasks that you need to carry out prior to flying your drone. Some of the things include making sure your batteries are well charged, that you have all of the equipment you need during the fight amongst other things. All this really helps to avoid things like crashing your drone.

★Take care of your Batteries

Taking care of your batteries is very crucial for a number of different reasons.

The batteries are the ones that power up your drone when flying hence it is very important to take good care of them. Some of the tips on taking care of your batteries include:

1. Flying with only fully charged batteries

2. Removing the battery from your drone when it is not in use.

3. Do not keep your battery on the charger for longer than 2 days

4. Do not fully drain your batteries

5. Avoid using a damaged battery on your drone.

6. Also, make a conscious effort to handle your batteries with great care!

★Keep your software updated

It is very important to check if your software is well up to date before flying. This is very important because if the firmware update is not updated, you may not be able to even fly your aircraft. Therefore, it is crucial that the firmware update is always up to date.

★Make sure your propellers are in good condition

Propellers play an important and crucial role when flying your drone. They are the ones that play an important role in making sure your drone is safe in the air. Therefore, if your propeller is cracked, chipped, or no longer sharp, it is best to replace it. Again, it is important to just carry a spare pair of propellers just to be safe.

★Keep your motors clean

Keeping the motors clean and well maintained is important. After flying your drone, you can remove the propellers and dust off or blow off any debris that could have been caught in between or inside the motors. Also, prevent your motors from over-heating or wearing out as this affects the flight.

★Fly your drone in good conditions

It is very important to fly your drone in good conditions, these are good weather conditions! The recommended flying weather conditions are usually stated in the drone's user manual. Generally, there are certain weather conditions that one cannot fly in. These include heavy snow, fog, and heavy rains. Flying in these bad conditions might definitely result in the damage of your drone.

It's also always good to keep some spares for your aircraft. These include a spare set of propellers, spare batteries, a soldering iron, and also a small toolkit for that good maintenance.

Feel free to comment below in the comments section or if you have any questions about this article!

Questions:

1) What does pre-flight check refer to?

2) Can the battery be kept in the drone for a long time when it is not in use?

3）What's the main role of propellers when flying a drone?

4）How to clean the motors after flying the drone?

5）In which weather conditions are the drone not allowed to fly?

【Project Evaluation】

Work activities	Weightage	Score
Reading and translation	4%	
Preparation and Clearance	5%	
Troubleshooting the UAV airframe	10%	
Repairing GCS	15%	
Maintaining RC	12%	
Maintaining UAV batteries	15%	
Repairing spring system of plant protection UAVs	15%	
Routine maintenance of UAVs	18%	
Attitude and attendance	6%	

Grade：□ Excellence □ Good marks □ Medium level □ Pass

【Project Conclusions】 This project mainly introduces the repair and maintenance of UAVs. Based on the introduction to common faults, the causes of the problems are analyzed, and then the solutions are found out, including the troubleshooting of airframe, GCS, battery, RC, minor repairs of spraying system of plant protection UAVs, and routine maintenance of UAVs.

Project 6　Mission Planning

【Description】 This project aims to create missions that will run when the drone switches to AUTO mode. You need a laptop, telemetry radios, batteries and a drone.

【Analysis】 UAV mission planning refers to the establishment of flight routes and task assignments for UAVs based on the tasks, the numbers of UAVs, and the types of mission payload.

The main goal of mission planning is planning one or more optimal or suboptimal trajectories from the starting point to the target point for UAV, which comprehensively considers the UAV's constraints based on terrain information and environmental conditions, such as performance, arrival time, energy consumption, threat, flight area and so on, ensuring that the UAV completes the flight mission efficiently and satisfactorily, and returns to home safely.

Generally, we use the ground control station to do the mission planning. There are at least ten different ground control stations. On desktop, there are DJI Ground station, ZERO-TECH Ground station, Mission Planner, APM Planner 2, MAVProxy, QGroundControl and UgCS. For Tablet/Smartphone, there are Tower (DroidPlanner 3), MAVPilot, AndroPilot and SidePilot that can be used to communicate with ArduPilot.

The decision to select a particular GSC often depends on your vehicle and preferred computing platform:

Ready-to-fly users may prefer the portability and ease of use of Tower, or another GCS running on a tablet or phone.

DIY/Kit users and developers often have to access configuration and analysis tools, and would therefore need (at least initially) Mission Planner, APM Planner 2 or another more full-featured GCS.

Here, we take the Mission Planner for example to illustrate the basic operations of mission planning for UAV.

【Related knowledge and skills】 UAV mission payload planning; UAV Laws and Regulations.

Read and translate some long and difficult sentences; know how to use Telemetry Radios.

Mission 1　The Basic Operations of Mission Planner

【Objective】Master the basic operations of Mission Planner.

【Analysis】Mission Planner is a full-featured ground station for the Plane，Copter and Rover. It is compatible with Windows only. Mission Planner can be used as a configuration utility or as a dynamic control supplement for your autonomous vehicle. With Mission Planner，you can plan，save and load autonomous missions into you autopilot with simple point-and-click way-point entry on Google or other maps.

【Knowledge preparation】UAV flight modes；aeronautical meteorology.

6.1.1　Mission Planner Overview

Mission Planner is a full-featured ground station application for Plane，Copter and Rover (see Fig. 6-1-1).

Fig. 6 - 1 - 1　Mission Planner

It is compatible with Windows only. Mission Planner can be used as a configuration utility or as a dynamic control supplement for your autonomous vehicle. Here are just a few things you can do with Mission Planner：

(1)Load the firmware into the autopilot board that controls your vehicle.

(2)Setup，configure，and tune your vehicle for optimum performance.

(3)Plan，save and load autonomous missions into you autopilot with simple point-and-click way-point entry on Google or other maps.

(4)Download and analyze mission logs created by your autopilot.

(5)Interface with a PC flight simulator to create a full hardware-in-the-loop UAV simulator.

(6) With appropriate telemetry hardware you can monitor your vehicle's status while in operation, analyze the telemetry logs, and operate your vehicle in FPV (first person view).

6.1.2　Connect Mission Planner to Autopilot

1.Setting up the connection

To establish a connection, you must first choose the communication method/channel you want to use, and then set up the physical hardware and Windows device drivers. You can connect the PC and autopilot using USB cables(see Fig. 6 - 1 - 2), Telemetry Radios(see Fig. 6 - 1 - 3), Bluetooth, IP connections etc.

The driver for your connection hardware must be present on Windows as this makes your connection's COM port and default data rate available to Mission Planner.

Fig. 6 - 1 - 2　Pixhawk USB connection

Fig. 6 - 1 - 3　Connection using SiK radio

On Mission Planner, the connection and data rate are set up using the drop-down boxes(see Fig. 6 - 1 - 4) in the upper right portion of the screen.

Fig. 6 - 1 - 4　COM port

Once you've attached the USB or Telemetry Radio, Windows will automatically assign your autopilot a COM port number, and that will show in the drop-down menu (the actual number does not matter). The appropriate data rate for the connection is also set (typically the USB connection data rate is 115,200 and the radio connection rate is 57,600).

Select the desired port and data rate and then press the Connect button to connect to the autopilot. After connecting Mission Planner will download parameters from the autopilot and the button will change to Disconnect as shown in Fig. 6 - 1 - 5.

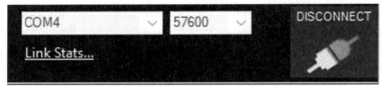

Fig. 6 - 1 - 5   Disconnect the port

2. Troubleshooting

If Mission Planner is unable to connect:

(1)Check that the correct baud rate is used for the selected method (115,200 on USB or 57,600 on Radio/Telemetry).

(2)If attaching via USB, be sure that a few seconds after power up have passed before attempting to connect. If you attempted to connect during the bootloader initialization time, Windows may get the wrong USB information. Connection attempts after this may require that the USB connection be unplugged and re-plugged, then wait for bootloader to enter the main code (a few seconds), then attempt the connection. Occasionally, MP must be restarted if an attempt to connect is made while in the bootloader initialization period.

(3)If using a COM port on Windows, check that the connection's COM port exists in the Windows Device Manager's list of serial ports.

(4)If your autopilot has an F7 or H7 processor and has CAN ports, then see the section below, Troubleshooting Composite Connections.

(5)If using a USB port, try a different physical USB port.

You should also ensure that the autopilot controller board has appropriate ArduPilot firmware installed and has booted correctly (on Pixhawk there are usefulLEDs and Sounds which can tell you the state of the autopilot).

If using a remote link (not USB)and Mission Planner connects, but does not download parameters or you cannot get commands, like mode changes acted upon, then the autopilot probably has Signing turned on. See MAVLink2 Signing.

6.1.3   Planning a Mission with Waypoints and Events

In this section, you will learn generic waypoint setup for all types of vehicles.

1. Setting the home position

The first necessary step is to set the home position, ensuring that UAV can automatically return to home and land when the UAV is in emergency situation, such as low battery, losing radio control signal, toggling to the RTL mode and so on. For Copter and Plane, the home position is set as the location where the vehicle was armed. This means the RTL mode is triggered, it will return to the location where it was armed, so arm your vehicle in the location you want it to return to, or use a rally point to setup an alternative return point.

2. Instructions

Fig. 6 – 1 – 6 shows that a Copter mission starts with an auto takeoff to 20 meters altitude; then goes to WP 2 rising to 100 meters altitude on the way and waits 10 seconds; then the craft will proceed to WP 3 (descending to 50 meters altitude on the way) and returns to launch. After reaching the launch position, the craft will land. The mission assumes that the launch position is set at the home position.

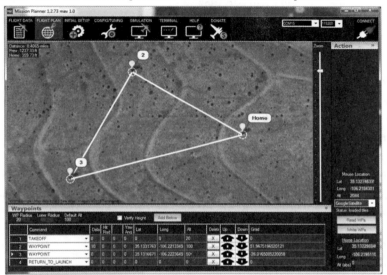

Fig. 6 – 1 – 6   Mission planning example

Now, you've got the potential! Just give it a shot!

You can enter waypoints and other commands. In the dropdown menus on each row, select the command you want. The column heading will change to show you what data that command requires. Lat and Lon can be entered by clicking on the map. Altitude is relative to your launch altitude or home position, so if you set 100m, for example, it will fly 100m above you.

In Fig. 6 – 1 – 6, "Default Alt" is the default altitude when entering new waypoints. ArduPilot documentation uses the word "altitude" often and in many different ways and contexts. Not all "altitudes" mean the same thing, and a good understanding of what altitude means in any given context is important to understand expected operation and behavior. Fig. 6 – 1 – 7 lists different "altitudes" and shows their differences and relations.

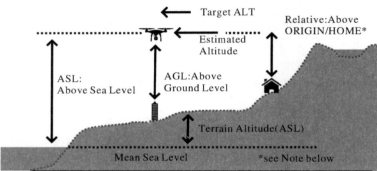

Fig. 6 - 1 - 7　ALT in different contexts

Note：ORIGIN and HOME are set during ground initialization from the GPS and are usually the same location，but the user may move the HOME location during the flight via Mission Planner (for a different RTL point)，but the Origin always remains the same unless intentionally reset.

　"Verify height" means that the Mission Planner will use Google Earth topology data to adjust your desired altitude at each waypoint to reflect the height of the ground beneath. So，if your waypoint is on a hill，if this option is selected，the Mission Planner will increase your ALT setting by the height of the hill. This is a good way to make sure you don't crash into mountains!

Once you are done with your mission，select "Write" and it will be sent to APM and saved in EEPROM. You can confirm that it's as you wanted by selecting "Read".

You can save multiple mission files to your local hard drive by selecting "Save WP File" or read in files with "Load WP File" in the right-click menu (see Fig. 6 - 1 - 8).

Fig. 6 - 1 - 8　Save or read mission file

3. Tips

Prefetch: You can cache map data so you don't need Internet access at the field. Click the Prefetch button, and hold down Alt to draw a box to download the selected imagery of a location.

Grid: This allows you to draw a polygon (right click) and automatically create waypoints over the selected area. Note that it does not do "island detection", which means if you have a big polygon and a little one inside of that, the little one will not be excluded from the big one (see this for more). Also, in the case of any polygon that partially doubles backs on itself (like the letter U), the open area in the center will be included as part of the flyover.

You can measure the distance between waypoints by right-clicking at one end and selecting Measure Distance. Then right-click on the other end and select Measure Distance again. A dialog box will open with the distance between the two points.

4. Auto grid

You can also have the Mission Planner create a mission for you, which is useful for function like mapping missions, where the aircraft should just go back and forth in a "lawnmower" pattern over an area to collect photographs.

To do this, in the right-click menu select Polygon and draw a box around the area you want to map. Then select Auto WP Grid. Follow the dialog box process to select altitude and spacing. The Mission Planner will then generate a mission that looks something like Fig. 6 - 1 - 9.

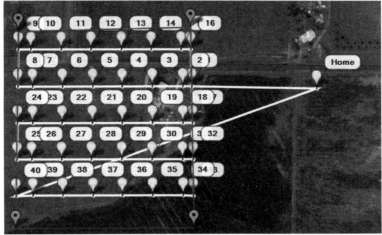

Fig. 6 - 1 - 9　Mission Planner auto-generated grid

5. Mission commands

Mission Planner provides a filtered list of the commands appropriate for the current vehicle type, and adds column headings for the parameters that need user-supplied values(see Fig. 6 - 1 - 10). These include navigation commands to travel to waypoints and loiter in the vicinity，DO commands to execute specific actions (for example taking pictures)，and condition commands that can control when DO commands are able to run.

Fig. 6 – 1 – 10　Mission command

Example：LOITER_TURNS command with headings for number of turns，direction，and location to loiter around.

The full set of mission commands supported by all ArduPilot platforms are listed inMAVLink Mission Command Messages (MAV_CMD). This includes the full name of each command (as defined in the protocol definition)，information about which parameters are supported，and also the corresponding Mission Planner column headings.

Note：

Mission Planner uses a cut-down version of the full command name.

For example，commands like MAV_CMD_NAV_WAYPOINT，MAV_CMD_ CONDITION_DISTANCE，MAV_CMD_DO_SET_SERVO are listed in MP as WAYPOINT，CONDITION_DISTANCE and DO_SET_SERVO respectively.

Words & Phrases

telemetry radio [təˈlemətri] [ˈreɪdiəʊ] 数传电台

trajectory [trəˈdʒektəri] *n.* 航线

waypoint [ˈweɪpɔɪnt] *n.* 航路点

dropdown [drɒpˈdaʊn] *n.* 下拉式(列表)，下拉式(菜单)

polygon [ˈpɒlɪɡən] *n.* 多边形

back and forth [bækənd fɔːθ] *adv.* 来回地

lawnmower [ˈlɔːnməʊə(r)] *n.* 割草机

arbitrary [ˈɑːbɪtrəri] *adj.* 任意的

Exercises and Thinking

1. Translate the following sentences.

1) It is compatible with Windows only. Mission Planner can be used as a configuration utility or as a dynamic control supplement for your autonomous vehicle.

2) If attaching via USB，be sure that a few seconds after power up have passed before attempting to connect.

3) The first necessary step is to set the home position，ensuring that UAV can automatically return to home and land when the UAV is in emergency situation，such as low battery，losing radio control signal，toggling to the RTL mode and so on.

2. Answer the questions in your own words.

1）Describe the full process of mission planning in brief.

2）How to set up the connection between Mission Planner and Pixhawk?

3）What is the home position? How to set the home position?

Mission 2　Plan Advanced Missions for Fixed-wing UAV

6.2.1　Basic Operation

In this section, I will show you basic functionalities of "flight plan" window (see Fig. 6 – 2 – 1), and you need a little bit of basics for doing some more advanced stuff. So, you can see the aerial imagery in the flight data screen that is a little bit bigger display here.

Fig. 6 – 2 – 1　Flight plan screen

In the flight data screen, you can see little green "H", which indicates the home position. The home position is actually gonna be reset when the aircraft is armed. Home position we see here is somewhat arbitrary which means that you can change the home position. Setting your home location to the current location is easy, just right click and select "Set Home Here", and it will set your home location to the current coordinates.

Now let's start to create a flight plan, you can click anywhere in the map which bring up the waypoint menu and create Waypoint 1 with an altitude of 150m. When you zoom in, you will see a dotted circle(see Fig. 6 – 2 – 2) which represent the radius of waypoint is 40m. The radius means that, as the aircraft is flying from home to point one, it will determine that this waypoint has been completed and it will start flying to waypoint two when it reaches this radius. So, the smaller the radius, the closer it will get to the waypoint before it turns, and the farther away the sooner it will turn. This is helpful in really windy situations where the aircraft may drift off track.

Fig. 6 - 2 - 2　Waypoints and circles

We're still able to complete the waypoint without circling back around. It's also useful because this is sort of almost predictive ability to turn and cut this corner and stay closer in line with the flight plan. we can create a few more way points.

And so, each of these waypoints you can see, have now shown up down here and we can get that little drag handle making the waypoint menu a little bigger so that we can see all of our waypoints, so we have Command which in this case is "waypoint", if you click that, you'll see a lot of other options(see Fig. 6 - 2 - 3), such as "LAND" "TAKEOFF""DO_SET_SERVO" and so on.

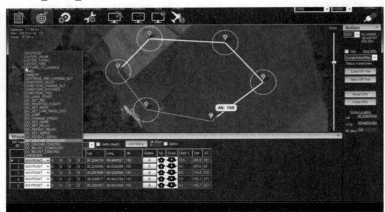

Fig. 6 - 2 - 3　Change the attribute of point

We also have a latitude, longitude and altitude of the waypoint. We have buttons right here to change the order of them with clicking "Up" and "down" arrow.

we have a gradient in percentage which is indicating the steepness of a climb or descent between two waypoints. You can change the altitude of waypoint 2 and observe how the percentage of climb change. For example, if we change this to two hundred and we can see that we have a 24.4% climb to get to a point 2 and 28.7% descent to get to a point 3(see Fig. 6 - 2 - 4). That will become important later in our flight planning because we want to keep those values as low as we practically can. Because steep climbs and steep descents are not good for the efficiency of our aircraft.

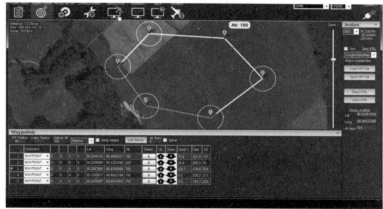

Fig. 6 - 2 - 4   The steepness between two waypoints

Fig. 6 - 2 - 5 shows that you can change the map source, we see lots of map options from dropdown. We can select any of these, such as GoogleMap, GoogleSatelliteMap, GoogleHybridMap, GoogleTerrainMap and so on.

Fig. 6 - 2 - 5   Select the map source

When you select GoogleTerrainMap, you can see there's terrain relief added in here, now I typically use hybrid because it gives me the best ability to discern where the flight plan is.

In the upper right, we have a latitude, longitude and elevation meters. This elevation in meters is terrain elevation based on GoogleTerrain data which is derived from the shuttle radar topography mission. so, it's a ninety-meter elevation dataset, so it's not super accurate, but it's pretty good for all of our flight planning purposes. so if we, for example, want to know how much terrain changes in elevation across the ground here, we can hover over home, and in the upper right we see that it's 226.6m above sea level. That's not gonna change a whole lot as we go out toward the river, because this is a nice flat field. When we drop down to the river, we still don't have much of elevation change, this is all flood plain. Then, if we go back up the hill, you'll see that we're climbing up to 260 meters, and this is important to us for making flight plans if we want to follow terrain, we have to take into account that terrain height change through there, if you're more comfortable working in UTM, you can click this dropdown right here (See Fig. 6 - 2 - 6). When you select UTM, you get a zone identifier, x, y and still the same elevation data.

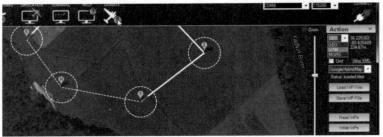

Fig. 6 - 2 - 6   UTM

And finally, one last thing that we have to do pretty much every time before flying, unless you like using a lot of cellular data to download aerial imagery, is prefetch image data sets. If you wanna go out and fly this flight plan, we want this imagery to be present behind here so that we can have an idea for where the aircraft is and if we were to go out and fly, we wouldn't have access to the Internet, we wouldn't have access to this Google imagery. And I would just have a blank screen behind this flight plan(see Fig. 6 - 2 - 7). So, we can hold down the ALT key, click and drag and we get a blue box.

Fig. 6 - 2 - 7   Select the area of image data sets

And then let go off the ALT key and right click, scroll down to the map tool and then click prefetch(see Fig. 6 - 2 - 8).

Then you're asked if you're ready to rip the imagery, just select yes and a series of these Windows will pop up prefetching symmetry. Typically, when it reaches zoom level 18, you can get much larger areas than this and because of that, this can become very time consuming because as you see the number of tiles that we have to fetch, increases quite a lot between each zoom layer. So, once I get to a sufficient zoom layer (usually zoom sixteen), we can just hit the escape key and make these go away just repeatedly hit the escape key until they stop showing up. Because blowing about zoom sixteen, you really don't get any increased image quality, so just skip out.

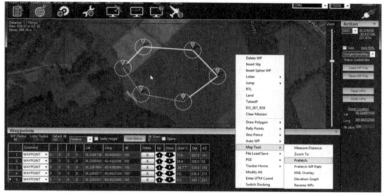

Fig. 6 – 2 – 8　Prefetch image data sets

One last little bit of helpful information for our flight planning appears at the upper left, we have "distance", "previous" and "home". Previous isn't particularly useful, but distance and home really are useful. So "home" is the distance in meters of the cursor from the home position. "distance" is the distance of the flight plan. So that is really helpful to us for creating flight plans, once you have a good idea for how far your aircraft can fly on a single battery, having this distance right here will be really helpful to you which will get the most out of every flight in making flight plans.

6.1.5　Advanced Mission Planning for Fixed-wing UAV

In this section, we will start to get into plans that will actually be useful to you in mapping, surveying, patrolling areas. So specifically, we're gonna talk about takeoff, landing, creating mapping grids, and then saving and loading flight plans once we have created them.

So, you can see the flight plan appears in the window (see Fig. 6 – 2 – 9), you have the same flight plan from last section and what we're gonna do is we're going to add a takeoff and landing point. In here we can actually fly this flight plan all in auto.

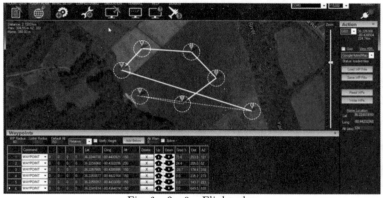

Fig. 6 – 2 – 9　Flight plan

So, first of all, I will right click and click "takeoff", that's gonna create a takeoff waypoint, it's gonna ask us for takeoff altitude. This value is the altitude at which the takeoff waypoint is considered complete. So, if we were to leave this at ten, the plane would reach ten meters above the ground and it would start flying to waypoint 1. The important thing is that this altitude should be high enough that we are clear of any terrain or trees before the aircraft starts to turn. So, if you're taking off from a relatively tight area surrounded by trees, your value should be well above any of those trees. I typically set it to sixty just to be safe so well above anything that we can turn freely without any terrain worries. You could set this higher but the problem with that is the climb during takeoff is quite inefficient. So, if you set this to something like 150, your takeoff will take you all the way to a cruising altitude, but it will do so very steeply and burn quite a lot of your available battery getting there. So, click OK, and you need to set a takeoff pitch for Castro aircraft, I typically use 25 to 30 degrees here, which is the maximum pitch of the aircraft during takeoff, so once the aircraft reaches cruise speed, it will climb at this angle. So, with our launch system, it is gonna be a cruise speed pretty much right at the end of the rails. Once the motor kicks on, it's going to nose up to this pitch and climb to the takeoff altitude at this pitch, so 25 degrees works well for us, if you have some obstacles that you need to clear, 30 may work better for you, 35 can even work. And again, the steeper the pitch value is, the less efficient your climb is gonna be. Click OK, and now we see down here at the bottom (see Fig. 6 - 2 - 10). For TAKEOFF command, we set pitch angle to 25, and you'll notice that the normally latitude and longitude columns are zero, this is because takeoff does not have a position. So, you could be anywhere on earth, load this flight plan click and take off regardless of where the flight plan is.

Fig. 6 - 2 - 10　Takeoff settings

You can get around this by creating the takeoff point before you create the flight plan or if we already have one, we can simply just click these up arrows to bring the "takeoff" all the way up to the top. And you will notice waypoint 1 becomes waypoint 2. So, what will happen? Will we launch the aircraft from near home? It will reach sixty meters and once it reaches sixty meters, it will turn toward waypoint 2 and continue around the flight plan.

Now we need to get back down on the ground, so we can just right click where we want it to land and click on "land", we get this landing waypoint position and altitude, and what would happen here is that the plane would retrieve waypoint 7 (see Fig. 6 - 2 - 11) and turn around to fly along this line and land right here.

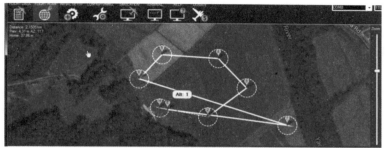

Fig. 6 - 2 - 11   Set landing waypoint

The problem with that is I don't necessarily want to land in a straight line with the final waypoint, this is sort of landing toward our parking lot and the trees, and we don't really want to do that if we overshoot or undershoot, we are not gonna be on the runway. So, I can delete waypoint 8, and before I create "takeoff", I will create an approach path. So, what we will do is we will create a series of way points, which will bring us a round and put us on a nice approach (see Fig. 6 - 2 - 12) to a landing here along the runway.

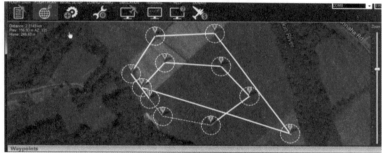

Fig. 6 - 2 - 12   Round approach path

So, you can see that we have these two waypoints (10 and 11) down here, and then we can click "land" right around home (see Fig. 6 - 2 - 13).

Fig. 6 - 2 - 13   Waypoint 12

So now we reach waypoint 7, turn left and come around this circuit. It's important to note that you won't land into the wind. So, if the wind is coming from the direction with waypoint 10 to 12, we don't want to land this way, we would want to make that circuit go around the other way and land towards the northwest. The aircraft flies pretty fast even there is no wind, so it's definitely not helpful to try to land down wind. The final thing that we need to do to make it work is to change the altitude of this approach path because we want it to slowly step down. And I'm gonna change waypoint eleven to fifteen meters, it will be fifteen meters and then it will cut the motor off and glide down to land. Then we really just want to go back and keep these descents relatively small, so we set the altitude of waypoint 10 to 35 so that we get a fairly shallow descent.

Other waypoints settings are shown in Fig. 6 – 2 – 14. So, we can see that we have a descent from 150 to 65 which is nice and gentle. Then we descend to 65. We have our landing point at one.

Fig. 6 – 2 – 14   Waypoint settings

So, this flight plan is ready to go, if we upload this to an aircraft and put the aircraft in auto and launch it. It would fly this plan, come back and land. Now we have a flight plan that is of some utility to us. We probably want to save this, so you may have flight plans that you use frequently and repeatedly, or you may have flight plans that put you on an approach path. Make a flight plan that is just an approach and landing for a flying site that uses frequently that you don't have to make an approach pattern every time, you can just load your approach, so the way that we save a waypoint file is we just go over the map and click "save wp file". And the default mission planner takes us to its program files folder, but you could save it anywhere on the computer. When you have this file saved, you can clear mission by right clicking "clear mission", you just left with our home point and can load it back up.

So now we will look at making a map grid. The first step to make any flight plan is adding a takeoff waypoint with altitude of 60 and pitch angle 25. And then we want to create a grid, so for this, I'm gonna map along this line down to where these fields change. Firstly, I'll go to the first corner of where we want to map, right click, scroll down to "draw polygon" and click "add polygon" notifying you that you will be in polygon mode until you clear the polygon, that's OK, and we get this little red icon. That is a corner on our polygon, so we can click to give ourselves three more corners. you can see that we create this red box (see Fig. 6 – 2 – 15), now we can have an arbitrary number of polygon points, so if we want to, we can click out somewhere and expand that out to a pentagon (see Fig. 6 – 2 – 16). There, if we accidentally did that, we could right click on that point and just like with the waypoint, we can click "delete wp", and that point would go away.

Fig. 6 - 2 - 15　Polygon

Fig. 6 - 2 - 16　Pentagon

Okay, now we have the area we would like to map, we want to right click, scroll down to "auto wp" which, as you might have guessed, is a way to automatically create waypoints. And then we'll go down to "survey (grid)". You can use survey gridv2 or simple grid, I don't really like either of them, "survey grid" gives us most powerful sort of user interface to create the grids. Click there, and other window pops up for us, there's a lot of information shown here (see Fig. 6 - 2 - 17). First of all, we see waypoints and flight path which create the grid. We have a lot of information about that flight which we'll talk about in a minute, but first of all, "area" is the area inside the red box in square meters. The "distance" is the distance of the grid flight plan, not the total flight plan, so if you have three kilometers that you have to fly out to this point and then fly the grid and fly back, this would be a total of nine kilometers, so this three is only for these grid lines. "Distance between images" is based on some information that we input, that is the distance between the time that the autopilot will trigger the camera. "Grand resolution" is approximately what your pixel size will be, and you can see the number of pictures needed to map this area, the number of strips, the number of transacts, the footprint of each image on the ground, distance between these transacts, the estimated flight time between image captures and turn diameter which is not really going to be useful to us at all.

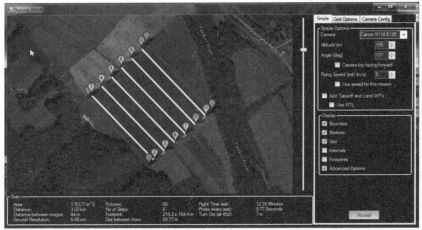

Fig. 6 - 2 - 17   Map grid

So, what we need to do to make this information and the flight plan correct is to set the options we need. The first and foremost is setting the camera, we typically fly with canon s110, you may fly with Nex5.

The focal length can influence the grid spacing, the spacing between the lines and the spacing between images because different cameras are gonna have different image footprints at this altitude. If you select 16mm, you can see that we need more lines (see Fig. 6 - 2 - 18)to cover this area because image footprint of the next five with the 16mm length is smaller.

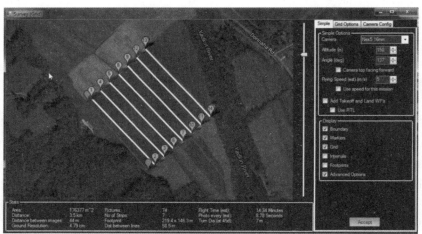

Fig. 6 - 2 - 18   7 Lines

Now we want to set the altitude to what we need in order to get the desired ground resolution, so for most purposes, 5. 5cm is more than enough. If you want better quality, you can change the altitude to 200. The tradeoff we get is quite a bit better resolution, but we have to fly quite a few more lines to cover that, because we have to maintain sufficient overlap. I would say fly as high as you can while getting the images you need.

Never fly lower than you need to, because you're just going to expend more energy to map the same area. The angle is automatically calculated by the algorithm that's creating this grid. And it is placing these lines along the longest axis of the polygon. If we don't like this direction, maybe we have a very strong wind coming out of the northwest. Then we don't wanna fly parallel with the wind, the reason being is flying upwind make the aircraft fly very slow which is not really a problem, but when we fly down, the aircraft is going to fly much faster. Then normally the airspeed will remain the same. As your ground speed is your airspeed plus the wind speed. If you're flying down with strong wind, your ground speed could be too high and you will actually fly faster than the camera can take images. So let's say we have a strong wind of the northwest, we can change this by ninety degrees (see Fig. 6 – 2 – 19). And now we're flying perpendicular to the wind, so we're gonna fly slower along all of these lines, but our speed will be consistent regardless of which direction in the grid we're flying.

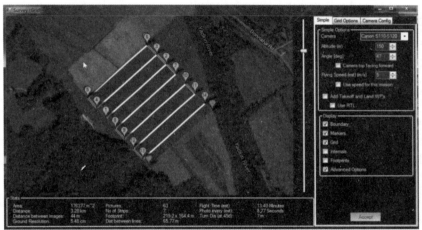

Fig. 6 – 2 – 19　Change the angle of grid

"Camera top facing forward" is important for the Castro aircraft. The camera top is not facing forward, so we don't want this check. If you're flying something like skywalker X8, we typically mount the camera top facing forward, you need to check this box, and that's actually nice because it allows you to have fewer lines. As the short axis of your images is now flying along transact, you have to take images more frequently. When I uncheck this, that "Photo every" goes up quite a bit. Finally, we want to change our flying speed, this actually isn't going to change anything about the grid when the aircraft behave is just gonna make this information more accurate. So, I'm going to set this to 18 which is our crew speed for the Castro in meters per second and now we see that if we're flying at eighteen meters per second, we need to take photos every 2. 44 seconds (see Fig. 6 – 2 – 20). The flight time for this grid will only be three minutes and forty-eight seconds, which like I said is a very short flight for us.

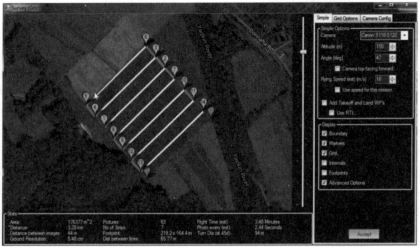

Fig. 6 – 2 – 20　Change flying speed

Okay, there are a few other things we can do. For example, we can go to grid options. The overlap and sidelap are important to have set correctly so that structure from motion software can work with this eighty percent and sixty percent respectively. Starting point in this case is set to home and you can see that waypoint 1 is right beside where the home point is. We could change this to top left and now we're starting up at one place and ending down at the other place. This can be useful in sort of making the most efficient flight plans, because you'll notice if we start from home, we fly six waypoints and we have to traverse back across this area to point 7, it's not the most efficient flight plan. So, you can play around with this to get what you think looks like the best option. There is none of the other stuff in grid options we need to worry about. In the camera configuration, if you have Sony Nex, but you have the 20mm lens that doesn't show up in list, you could click on the sixteen and then you can go to camera config and change the focal length to twenty, your other parameters of the camera like the sensor size can still be correct. We found there are more lines than sixteen millimeters which makes sense, because our footprint will be smaller with the longer length, you can also save it so that in future you'll have it showing up as a safety camera.

For now, I will go back to the s110, because like I said that is what we typically use. Now, if we click "accept", we see that this grid has popped up on screen and it looks like I left starting point at home which is fine for now. You can see waypoint 2 is near the home, because we have "takeoff" being waypoint 1. Another important thing to note here is that you notice this Do_SET_CAM_TRIGG_DIST command (see Fig. 6 – 2 – 21), it comes after the first waypoint on the grid, then another one comes after the last waypoint on the grid, now what this is doing is once it reaches the first waypoint, it is activating the camera trigger, so it will trigger every 43. 84 and so on meters which is selected to give us the correct overlap between images.

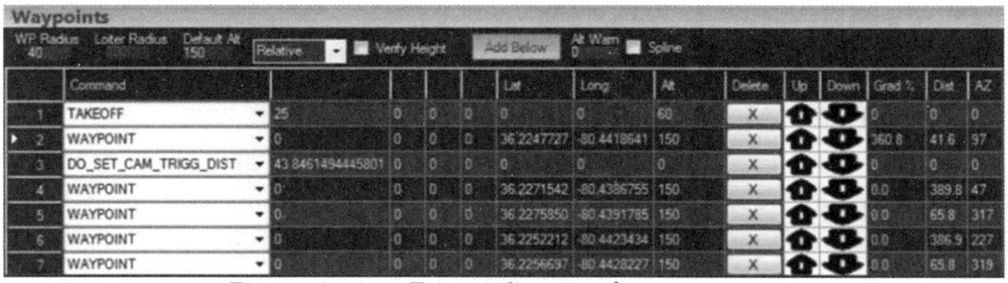

Fig. 6 – 2 – 21 Trigger distance after waypoint 2

Then at the last waypoint, we set the trigger distance to zero (see Fig. 6 – 2 – 22) which turns off automatic camera triggering so that we are not taking images while we are on approach, that can also be really important for camera like the s110 where we need the lens to retract before we land if we have that in the belly of plane.

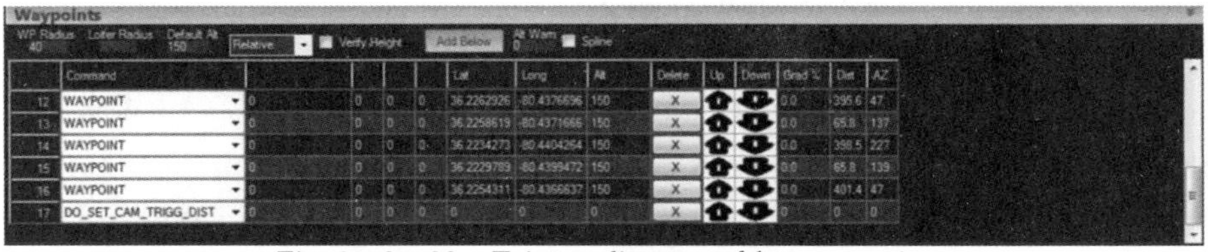

Fig. 6 – 2 – 22 Trigger distance of last waypoint

So, we have this grid, all that's left for us to do is to make an approach pattern. I wanna make waypoints surround home and landing, but if I click, you see that I get another polygon (see Fig. 6 – 2 – 23). That's because like that dialog that pops up said, we're still in polygon mode. So right click, scroll down to "draw polygon", click "clear polygon", and now the points go away, you can still see the red line. But if I move the map, they disappear.

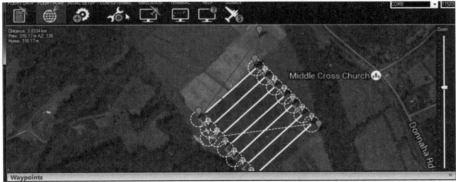

Fig. 6 – 2 – 23 Another polygon

And then when we click, we get waypoints just like normal. Zoom in and move these waypoints exactly to give us the proper approach to land. Right click to get landing points and then just like before, go back and set altitudes to give us an acceptable approach making everything is below fifteen degrees or fifteen percent descent, so we're good to go and have a flight plan that works for us.

One last feature that we need to look at when we're doing a grid flight plan is extending and retracting the camera lens. On our aircraft like the Castro, we typically use canon s110 when the lens is extended. It extends out of the bottom of the aircraft. it's really important that that lens is retracted when the aircraft lands. The way we do that is fairly simple. What we want to do is make a waypoint which will do that extending and retracting for us. What we do is extending lens immediately after takeoff and we will retract a few waypoints before landing, so we're sure that it is after the last pictures taken, but before we touch down.

So, I'll click on the takeoff point and click "add below". What that's gonna do is creating a waypoint at (0,0)with zero altitude. We don't want that but what we do want is to change this waypoint drop down box. And we want to change this to "DO_SET_SERVO". Click on "DO_SET_SERVO" and change the servo value to extend the camera lens because the camera is controlled by one of the servo outputs of the autopilot. We can see the title of the first column changes into servo number and PWM is the second column which is a signal value. You need to check the servo number that operating our camera and input the correct data. If we want to set this PWM value to 1,100. Now don't worry about what "1,100" means, knowing that it is the low value of the servo, the high value is 1,900. We just want to set the DO_SET_SERVO command for servo 5 to 1,100 which will extend lens right after takeoff. And now, we can also alternatively go to insert waypoint, in this case we want to insert the waypoint well after we stopped taking pictures, but well before we land. So we can insert a waypoint after waypoint 20 (see Fig. 6 - 2 - 24).

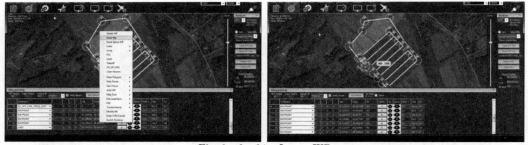

Fig. 6 - 2 - 24　Insert WP

Now we have waypoint 21, but all we need to do is changing that from waypoint to "DO_SET_SERVO" just like before. Change the serve number to 5 and PWM to high value with 1900 which will retract our camera. Other sets can be zero. Then you can see when we change this to "DO_SET_SERVO", that point 21 goes away, but you do see that we jump from 20 to 22. So, point 21 became invisible. It's really important to remember to insert these "DO_SET_SERVO" commands to extend and retract lens. If you do not do that, the camera lens will be down when the aircraft lands and it will damage or destroy the camera.

6.1.6　Conclusions

(1) There is a situation where the user will find that it will not connect to the obvious COM port in the Mission Planner dropdown box. This occurs when the user accidentally changes the protocol of whichever SERIALx port the Windows driver is using as the MAVLink COM port to something other than MAVLink. This can easily happen if the user takes an existing parameter file from a vehicle configuration used with a different autopilot that has the protocol changed.

(2) Mission Planner exposes the full subset of commands and parameters supported by ArduPilot, filtered to display just those relevant to the currently connected vehicle.

(3) Navigation commands are used to control the movement of the vehicle, including takeoff, moving to and around waypoints, and landing.

Words & Phrases

pentagon['pentəgən] *n.* 五边形

tradeoff ['treɪˌdɔf] *n.* 折中;权衡

footprint['fʊtprɪnt] *n.* 覆盖区

belly ['beli] *n.* 腹部

takeoff ['teɪˌkɔf] *n.* 起飞

be clear of [biːklɪə əv] 摆脱;除掉

Exercises and Thinking

1. Translate the following sentences.

1) Home position we see here is somewhat arbitrary which means that you can change the home position.

2) So, if you're taking off from a relatively tight area surrounded by trees, your value should be well above any of those trees.

3) One last feature that we need to look at when we're doing a grid flight plan is extending and retracting the camera lens.

4) What we do is extending lens immediately after takeoff and we will retract a few waypoints before landing, so we're sure that it is after the last pictures taken, but before we touch down.

2. Answer the questions in your own words.

1) When the UAV is conducting an automatic flying mission, and we want to change the altitude of waypoint that the UAV has not reached yet temporarily, what should we do?

2) If you want to do a survey on the growth of farmland crop with UAV, what should you do to complete this task effectively?

3) How to plan a mission with waypoints and events?

【Project Evaluation】

Work activities	Weightage	Score
Reading and translation	4%	
Preparation and clearance	5%	
Connect Mission Planner to AutoPilot	10%	
Planning a mission with waypoints and events	20%	
Basic operation	25%	
Advanced mission planning for fixed-wing UAV	30%	
Attitude and attendance	6%	

Grade：□ Excellence   □ Good marks   □ Medium level   □ Pass

【Project Conclusions】 This project introduces operations of mission planning with Mission Planner. It starts with the connection between GCS and UAV, and provides troubleshooting methods when we cannot connect to the COM port; Then, follow the project instructions, we can plan a mission with waypoints and events, such as setting home position, generating auto grid, setting mission commands and so on; After we know exactly what is mission planning through step 2, we learn basic functionalities of "flight plan" window which is very useful for creating flight plans; Finally, we take advanced mission planning for example, which helps us thoroughly study the whole process of planning a mission for drone.

References

[1] Wikipedia. Unmanned Aerial Vehicle [EB/OL]. [2018 - 03 - 06]. https://en. wikipedia. org/wiki/Unmanned_ aerial_vehicle.

[2] Wikipedia. Quadcopter [EB/OL]. [2018 - 03 - 07]. https://en. wikipedia. org/wiki/Quadcopter.

[3] ArduPilot Dev Team. What You Need to Build a MultiCopter [EB/OL]. [2018 - 03 - 08]. https://ardupilot. org/copter/docs/.

[4] Futaba Corp. Futaba T14SG Instruction Manual [EB/OL]. [2018 - 03 - 08]. https:// futabausa. com/wp - content/uploads/2018/09/14SG. pdf.

[5] SkyWalker. Skywalker X - 8 Installation Manual [EB/OL]. [2022 - 02 - 21]. http:// skywalkermodel. com/en/76. html♯.

[6] Align Communications, Inc. Align 450Pro Instruction Manual [EB/OL]. [2022 - 12 - 21]. https://www. align. com. tw/index. php/agriculturalmulticopter - cn/.